T0139312

MEDICAL
INTELLIGENCE
UNIT

Somatostatin Analogs in Diagnostics and Therapy

Marek Pawlikowski, M.D., Ph.D.
Department of Neuroendocrinology
Medical University of Lodz
Lodz, Poland

LANDES BIOSCIENCE
AUSTIN, TEXAS
U.S.A.

SOMATOSTATIN ANALOGS IN DIAGNOSTICS AND THERAPY

Medical Intelligence Unit

Landes Bioscience

Please address all inquiries to the Publishers:
Landes Bioscience, 1002 West Avenue, Second Floor, Austin, Texas 78701 U.S.A.
Phone: 512/ 637 6050; Fax: 512/ 637 6079
www.landesbioscience.com

ISBN: 978-1-58706-223-0

While the authors, editors and publisher believe that drug selection and dosage and the specifications and usage of equipment and devices, as set forth in this book, are in accord with current recommendations and practice at the time of publication, they make no warranty, expressed or implied, with respect to material described in this book. In view of the ongoing research, equipment development, changes in governmental regulations and the rapid accumulation of information relating to the biomedical sciences, the reader is urged to carefully review and evaluate the information provided herein.

Library of Congress Cataloging-in-Publication Data

Library of Congress Cataloging-in-Publication Data

Somatostatin analogs in diagnostics and therapy / [edited by] Marek Pawlikowski.
  p. ; cm. -- (Medical intelligence unit)
 Includes bibliographical references and index.
 ISBN 978-1-58706-223-0
 1. Somatostatin. I. Pawlikowski, M. (Marek) II. Series: Medical intelligence unit (Unnumbered : 2003)
 [DNLM: 1. Somatostatin--analogs & derivatives. 2. Somatostatin--physiology. 3. Neuroendocrine Tumors--diagnosis. 4. Neuroendocrine Tumors--drug therapy. 5. Receptors, Somatostatin--physiology. WK 515 S6932 2007]
 QP572.S59S66693 2007
 612'.015756--dc22

                    2007007925

About the Editor...

MAREK PAWLIKOWSKI is Professor of Endocrinology in the Medical University of Lodz, Poland. In years 1975-2004 he was Director of the Institute of Endocrinology, Medical University of Lodz, and in years 1987-1990 President (Rector) of this medical school. His main research interest is neuroendocrinology and oncological endocrinology. He is a founder and the honorary President of the Polish Society of Neuroendocrinology, honorary member of the Polish Society of Endocrinology and honorary editor of Neuroendocrinology Letters.

CONTENTS

EDITOR

Marek Pawlikowski, M.D., Ph.D.
Department of Neuroendocrinology
Medical University of Lodz
Lodz, Poland
Email: pawlikowski.m@wp.pl
Chapters 1, 4, 8

CONTRIBUTORS

Marianna I. Bak
Department of Nuclear Medicine
Medical Academy of Warsaw
Warszawa, Poland
Chapter 7

Mª José Barahona
Department of Endocrinology
Hospital Sant Pau
Department of Medicine
Autonomous University of Barcelona
Barcelona, Spain
Chapter 5

Michael D. Culler
Endocrine Research
Biomeasure Incorporated/IPSEN
Milford, Massachusetts, U.S.A.
Email: michael.culler@ipsen.com
Chapter 3

Beata Kos-Kudla
Department of Pathophysiology
   and Endocrinology
Silesian Medical Academy
Zabrze, Poland
Email: beatakos@ka.onet.pl
Chapter 6

Leszek Królicki
Department of Nuclear Medicine
Medical Academy of Warsaw
Warszawa, Poland
Email: krolicki@amwaw.edu.pl
Chapter 7

Jolanta Kunikowska
Department of Nuclear Medicine
Medical Academy of Warsaw
Warszawa, Poland
Chapter 7

Gabriela Melen-Mucha
Department of Immunoendocrinology
Medical University of Lodz
Lodz, Poland
Email: g.m-mucha@wp.pl
Chapters 2, 8

Slawomir Mucha
Department of Clinical Endocrinology
Medical University of Lodz
Lodtz, Poland
Chapter 2

Nuria Sucunza
Department of Endocrinology
Hospital Sant Pau
Department of Medicine
Autonomous University of Barcelona
Barcelona, Spain
Chapter 5

Susan M. Webb
Department of Endocrinology
Hospital Sant Pau
Department of Medicine
Autonomous University of Barcelona
Barcelona, Spain
Email: swebb@santpau.es
Chapter 5

PREFACE

Somatostatin is a peptide isolated originally from the hypothalamus and considered as an inhibitor of growth hormone secretion. However, further studies have shown that the peptide was ubiquitously distributed and exerts a large spectrum of physiological effects, mostly of an inhibitory nature. The very short half-life of the native peptide limits its therapeutic application. To overcome this limitation the long-lasting modified molecules (analogs) of somatostatin were synthesized. The present book provides comprehensive data on the application of somatostatin analogs in diagnostics and treatment of diseases, mostly endocrine disorders and cancers.

Two chapters (1 by M. Pawlikowski and 2 by G. Melen-Mucha and S. Mucha) are devoted to the physiological background and report on the tissue distribution of somatostatin, its physiological effects, as well as on the tissue distribution and function of somatostatin receptors. Chapter 3 by M.D. Culler discusses the complexity of somatostatin analog action linked to the interaction of receptor subtypes as well as the heterodimerization of somatostatin and dopamine receptors. Chapter 4 by M. Pawlikowski describes somatostatin receptors in human tumors as revealed by ex vivo-in vitro approaches. Special attention was paid to somatostatin receptor immunohistochemistry as a simple and effective mode of their detection in surgical and biopsy samples. Chapters 5 by N. Sucunza et al and 6 by B. Kos-Kudla report on the most well-established therapeutic indications of somatostatin analogs: acromegaly and the so-called neuroendocrine/gastrointestinopancreatic tumors (NET/GEP), respectively. In Chapter 7 we can find the data on the detection of somatostatin receptors in vivo using radiolabeled somatostatin analogs and on the novel promising therapy of somatostatin receptor-bearing cancers—somatostatin receptor targeted radiotherapy. The last chapter of the book (Chapter 8 by G. Melen-Mucha and M. Pawlikowski) reviews the future perspectives of new potential therapeutic applications of somatostatin analogs. These new perspectives include, among others, the treatment of endocrine an non-endocrine cancers, especially with somatostatin analogs coupled to cytotoxic drugs, and the treatment of chronic inflammatory and autoimmune diseases.

Marek Pawlikowski, M.D., Ph.D.

CHAPTER 1

Physiology of Somatostatin

Marek Pawlikowski*

Abstract

Somatostatin (SST) was originally discovered as a hypothalamic peptide which inhibits growth hormone (GH) secretion from the pituitary gland. It appears in two molecular forms, composed of 14 or 28 amino-acid residues. Moreover, another family of peptides, called cortistatins (CST), was described. CST are encoded by a different gene, but they share partly the structure of SST and bind to SST receptors. Further studies revealed that SST is expressed not only in the hypothalamus but is widely distributed in central and peripheral nervous systems as well as in nonneural peripheral tissues, mainly in the gut. SST is now known to exert a large spectrum of functions, mostly of an inhibitory nature. It inhibits the secretion of hormones, like GH and thyrotropin (TSH), neuro-enterohormones like gastrin, cholecystokinin, vasoactive intestinal peptide (VIP), gastric inhibitory polypeptide (GIP), motilin, secretin, pancreatic polypeptide and glucagon-like peptides (GLP), and exocrine secretions. SST modulates the functions of nervous and immune systems and exerts a direct antiproliferative effect on cell and tissue growth. Because of a very short half-life (approx. 60 seconds) the therapeutic application of exogenous SST is limited. To overcome this obstacle, the long-acting analogs of SST were synthesized.

Introduction

Somatostatin (SST) was originally isolated from the hypothalamus and considered as hypothalamic neurohormone inhibiting the growth hormone (GH) secretion (somatotropin release inhibitng factor—SRIF). Its molecule was identified as a tetradecapetide.[1] Later another molecular form of SST was also found, composed of 28 amino-acid residues and included the tetradecapeptide sequence.[2] These molecular forms were called somatostatin-14 (SST-14) and somatostatin-28 (SST-28), respectively (Fig. 1). Both forms of somatostatin generate by the posttranslational processing of a larger molecule (composed of 116 amino-acid residues) called preprosomatostatin. Although SST was originally considered as hypothalamic hormone and its function was linked initially only to the control of GH secretion, it was early established that the hormone is widely distributed within the central nervous system as well as in peripheral tissues and possesses a wide spectrum of functions, including inhibition of various endocrine and exocrine secretions, neurotransmission, immunomodulation and antiproliferative effects.

SST exerts its biological effects acting via somatostatin receptors (sst). Five distinct sst rreceptor proteins (sst 1-5) were cloned. They are encoded by distinct genes located on different chromosomes. Moreover, one of the receptor subtypes (sst2) occurs in two splicing variants denoted as sst2A and sst2B. Both native SST peptides, SST-14 and SST-28 bind with high affinity to all subtypes of sst receptors. The detailed data on sst receptors are presented in Chapter 2 of this

*Marek Pawlikowski—Department of Neuroendocrinology, Chair of Endocrinology, Medical University of Lodz, Sterling str 3, 91-425 Lodz, Poland. Email: pawlikowski.m@wp. pl

Somatostatin Analogs in Diagnostics and Therapy, edited by Marek Pawlikowski.
©2007 Landes Bioscience.

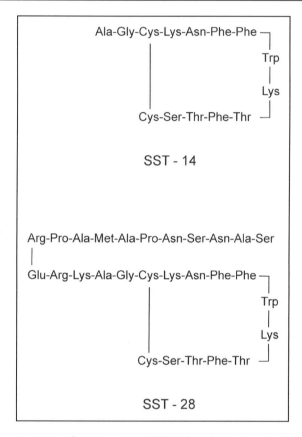

Figure 1. Primary structures of somatostatin-14(SST-14) and somatostatin-28 (SST-28).

book. The native SST peptides have a very short half-life time (approx. 60 seconds). For this reason the usefulness of exogenous SST in vivo either in experimental or in therapeutic purposes is limited. To overcome this difficulty, numerous long acting SST analogs were synthesized. These analogs—in contrast to the native peptides—markedly differ in their affinities to particular subtypes of sst receptors (see Chapter 3 of this book). SST-14 and SST-28 are not sole endogenous ligands for sst receptors. In nineties, another kind of endogenous peptides binding to sst receptors have been discovered and called cortistatins (CST).[3] CST are generated from the precursor molecule of 112 amino-acid residues called preprocortistatin. Preprocortistatin shares some structural similarities with SST precursor molecule, preprosomatostatin, but is encoded by a different gene. CSTs occurr in different molecular forms, composed of 13,14,17 or 29 amino-acid residues and their structures are similar to SST. For instance, CST-14 shares 11 amino-acid residues with SST-14. Because CSTs interact with all subtypes of sst receptors, they share their physiological actions with SST. However, besides sst receptors, CSTs bind also to GH secretogogue receptors (ghrelin receptors) and the so-called MrgX2 receptors. The latter is considered as specific CST receptor. For this reason, CSTs differ in some physiological effects from native SSTs.

Tissue Distribution of Somatostatin

Within the central nervous system, SST is detected mainly in the hypothalamus, but is also widely distributed outside this region. In the peripheral nervous system, SST is mainly localized in dorsal roots ganglia and in sensory nerves. Outside the central and peripheral nervous system SST is expressed mainly in the pancreas and gastrointestinal tract. In the pancreas, SST is localized in D-cells of pancreatic islets. In the gastrointestinal tract SST is mainly present in endocrine D-cells of the mucosal layer. However, in the small intestine an important part of SST is localized in the nerve cell bodies of the enteric and myenteric ganglia. The highest concentrations of SST were present in the stomach and duodenum, lower in the remaining part of the small intestine. In contrast, the colon is poor in SST.[4]

In the thyroid gland, SST is colocalized with calcitonin in parafollicular (C) cells.[5,6] SST immunoreactive cells were also detected in the human bone marrow.[7] The expression of SST was also found in human renal cortex and mesangial cells in the kidney.[8]

The rodent immune cells were shown to synthesize and secrete SST. SST mRNA was found in macrophages of the murine granuloma induced by *Schisostoma mansoni* infection. Contrary, T lymphocytes within the granuloma do not express SST but express sst receptors (of sst2 subtype). In rat both T and B lymphocytes are able to produce SST.[9] Contrary to rodents, in the human immune system SST mRNA can be detected only in the epithelial compartment of the thymus. On the other hand, CST mRNA was found in T and B lymphocytes, monocytes, macrophages and dendrite cells.[10] It was suggested that CST rather than SST is an endogenous ligand for sst receptors expressed by human immunocytes.

Functions of Somatostatin

Central and Peripheral Nervous System

SST acts in the central and peripheral nervous systems as neuromodulator and neurotransmitter. It modulates the neurotransmitter release increasing that of 5-hyroxytryptamine and dopamine and decreasing the release of GABA and norepinephrine. Administration of exogenous SST or its analog octreotide exerts several neuromodulatory effects like antinociceptive effects,[11] modulation of sleep[12] or inhibition of epileptoid seizures.[13]

Anterior Pituitary Gland

The classic function of SST is the inhibition growth hormone (GH) secretion. This inhibitory effect of SST is exerted on two levels: SST inhibits the release of growth hormone releasing hormone (GHRH) from the hypothalamus and acts also directly on pituitary somatotrophs, suppressing their secretory function. Besides of GH, SST is also a physiological inhibitor of thyrotropin (TSH) secretion.

Peripheral Endocrine Glands

In the thyroid gland SST inhibits the TSH-induced secretion of thyroid hormones[14,15] as well as TSH-stimulated proliferation of thyrocytes[16] SST analog octreotide was found to counteract experimental goitrogenesis induced by propylotiouracil.[17] In the adrenal gland, SST was found to suppress the aldosterone secretion, acting indirectly via inhibition of renin production in kidneys and directly on sst receptors in the adrenal zona glomerulosa.

Gastrointestinal System

In the pancreas SST suppresses both endocrine and exocrine secretions. It inhibits the secretion of insulin and glucagon from pancreatic islets, digestive enzymes and bicarbonate from the exocrine pancreas. In the liver, SST inhibits the bile production In the gut, SST inhibits exocrine secretion of gastric acid and pepsinogen as well as endocrine/paracrine secretions of the numerous peptide neuro-enterohormones like gastrin, cholecystokinin, vasoactive intestinal peptide (VIP), gastric inhibitory polypeptide (GIP), motilin, secretin, pancreatic polypeptide and glucagon-like peptides (GLP). The important effects of SST in the gut is the inhibition of gastrointestinal motility and the reduction of the splanchnic blood flow (for review see ref. 4).

The Immune System

The immunocytes express sst receptors and are targets for SST. The effects of exogenous native SST and/or SST analogs involve the modulation of lymphoid cells proliferation, and the production of immunoglobulins and cytokines. The extensive studies of SST effects on lymphocyte and lymphoid cell lines proliferation revealed inhibitory or biphasic effects of the hormone. The inhibitory effects were observed with low (<1 nM) whereas stimulatory with high concentrations of SST.[18,19] SST was found to inhibit the immunoglobulins production, particularly of IgG A.[20] The effects of SST and/or SST analogs on cytokine production are divergent, but mostly inhibitory.[21-24] The inhibition of the pro-inflammatory cytokines secretion is probably the main mechanism of the anti-inflammatory action of SST analog octreotide, which has been proved on different animal models of inflammation. like the murine granuloma induced by Schisostoma Mansoni infection,[25] carageen-induced inflammation in rat,[26] neurogenic inflammation of guinea pig bladder,[27] adjuvant arthritis in rat[28] and zymosan-induced earlobe inflammation in mouse.[29]

Because CST but not SST is expressed in the human immune system, it seems that CST rather than SST is an endogenous modulator of the immune functions in man acting via sst receptors (see above).

The Vascular System

SST-14 and some of its analogs were shown to inhibit angiogenesis in chorioallantoic membrane of chicken embryo.[30] This anti-angiogenic activity was further confirmed on other in vivo models. For instance, SST analog octreotide suppressed angiogenesis in rat cornea after chemical injury and FGF-induced neovascularization in the rat mesentery.[31] Octreotide inhibits also neovascularization in the estrogen-treated anterior pituitary gland [32] and in the propylotiouracil-treated rat thyroid.[17] The in vitro studies suggest that the angiosuppressive effects of SST and its analogs depend on the direct inhibitory action on endothelial cell proliferation. SST-14 and its analog octreotide was found to inhibit the proliferation of murine line of endothelial cells HeCa10.[33] It was also shown that SST analogs octreotide and SOM 230 suppressed the in vitro proliferation of another endothelial cell line, HUVEC (human umbilical vein endothelial cells).[34] In contrast, neither SST-14 nor octreotide inhibited the vascular endothelial cell growth factor (VEGF) secretion from HeCa10 cells.[33]

Direct Antiproliferative and Pro-Apoptotic Actions

It is well known that SST, besides its influence on the growth of tissues and organs mediated by GH inhibition, exerts also the direct antiproliferative effects at the cellular level acting via sst receptors. The early observation of the direct antimitogenic action of SST was done in our laboratory and concerned the TRH-stimulated mitotic activity of the rat anterior pituitary gland.[35] Besides the anterior pituitary gland, the direct antiproliferative effects of SST was demonstrated in gastric and intestinal mucosa, thyroid gland, exocrine pancreas, bone marrow and lymphocytes (for review of the early findings see ref. 36). It can be assumed that all tissues expressing sst receptors respond to SST by the inhibition of cell growth. The above concerns not only normal tissues, but numerous neoplastic tissues as well. The first observation of growth suppression of neoplastic cells exposed to SST in vitro was done by Mascardo and Sherline[37] in the early eighties. The observation concerned HeLa cells and gerbil fibroma cells. The list of tumoral cell lines of human or animal origin responding in vitro to SST or its analogs by growth inhibition is long and includes pituitary adenomas, thyroid, adrenal, neuroendocrine, gastric, colonic, pancreatic, hepatocellular, prostate and breast cancers and osteosarcomas. The molecular mechanisms of the antiproliferative action of SST are complex and include the activation of tyrosine phosphatases and serine-threonine phosphatases. The activation of tyrosine phosphatases leads to dephosphorylation of proteins which phosphorylation is stimulated by tyrosine kinases,[38] thereby inhibiting receptor tyrosine kinases-mediated signals and inhibition of receptor tyrosine kinases themselves.[39] It is well known that the latter mediate the mitogenic effects of numerous growth factors. In turn, serine-threonine phosphatases counteract the activity of serine-threonine kinases like MAP kinases[40] and Akt/PKB.[41] Recently, an additional mechanism has been proposed: the inhibition of telomerase, an enzyme closely linked to the attainment of immortality by cancer cells.[41] Another mechanism by which SST controls the excessive growth is the induction of apoptosis. The pro-apoptotic action of SST is mediated by sst3 receptor subtype[42] and involves such suppressor genes like p53[43] and Bax.[43,44]

Conclusions

Somatostatin (SST), originally discovered as a hypothalamic neurohormone which negatively regulates the growth hormone secretion from the pituitary, is widely distributed within the body and presents the wide spectrum of functions. The functions of SST are mostly inhibitory and include the suppression of divergent endocrine and exocrine secretions as well as tissue and cell growth. because of a very short half-life time of the native SST, the use of the synthetic long-acting molecules (SST analogs) is needed to take advantages of SST properties for therapeutic purposes.

References

1. Brazeau P, Vale W, Burgus R et al. Hypothalamic polypeptide that inhibits the secretion of immunoreactive pituitary growth hormone. Science 1973; 179:77-79.
2. Pradayrol L, Jornvall J, Mutt V et al. N-terminally extended somatostatin: The primary structure of somatostatin-28. FEBS Lett 1980; 109:55-59.
3. de Lecea L, Criado JR, Prospero-Garcia O et al. A cortical neuropeptide with neuronal depressant and sleep modulating properties. Nature 1996; 381:242-245.
4. Chiba T, Yamada T. Gut somatostatin. In: Walsh JH, Dockray GJ, eds. Gut Peptides: Biochemistry and Physiology. New York: Raven Press, 1994:123-145.
5. Van Noorden S, Polak JM, Pearse AGE. Single cellular origin of somaotstatin and calcitonin in the rat thyroid gland. Histochemistry 1977; 53:243-247.
6. Alumets J, Hakanson R, Lundqvist G et al. Ontogeny and ultrastructure of somatostatin and calcitonin in the thyroid gland of the rat. Cell Tissue Res 1980; 206:193-201.
7. Godlewski A. Calcitonin and somatosatin immunoreactive cells are present in human bone marrow and bone marrow cells are responsive to calcitonin and somatostatin. Exp Clin Endocrinol 1990; 96:219-233.
8. Turman MA, O'Dorisio TM, Apple CA et al. Somatostatin expression in human renal cortex and mesangial cells. Reg Peptides 1997; 68:15-21.
9. Aguila MC, Dees WL, Haensly W et al. Evidence that somatostatin is localized and synthesized in lymphoid organs. Proc Natl Acad Sci USA 1991; 88:11485-11489.
10. Dalm VA, van Hagen PM, van Koetsveld PM et al. Cortistatin rather than somatostatin as a potential endogenous ligand for somatostatin receptors in the human immune system. J Clin Endocrinol Metab 2003; 88:270-276.
11. Chrubasik J, Meynadier J, Scherpeerel P et al. The effect of epidural somatostatin on postoperative pain. Anesth Analg 1985; 64:1085-1088.
12. Hajdu I, Szentirmai E, Obal F et al. Different brain structures mediate drinking and sleep suppression elicted by the somatostatin analog, octreotide, in rats. Brain Res 2003; 994:115-123.
13. Vezzani A, Hoyer D. Brain somatostatin: A candidate to inhibitory role in seizures and epileptogenesis. Eur J Neurosci 1999; 11:3767-3776.
14. Loos U, Raptis S, Birk J et al. Inhibition of TSH-stimulated radioiodine turnover and release of T4 and T3 in vitro by somatostatin. Metabolism 1978; 27:1269-1273.
15. Ahren B, Ericsson M, Hedner P et al. Somatostatin inhibits thyroid hormone secretion induced by exogenous TSH in man. J Clin Endocrinol Metab 1978; 47:1156-1159.
16. Zerek-Melen G, Lewinski A, Pawlikowski M et al. Influence of somatostatin and epidermal growth factor (EGF) on the proliferation of follicular cells in the organ-cultured rat thyroid. Res Exp Med 1987; 187:415-421.
17. Pawlikowski M, Zielinski K, Slowinska-Klencka D et al. Effects of octreotide on propylotiouracil-induced goiter in rats: A quantitative evaluation. Histol Histopathol 1998; 13:679-682.
18. Pawlikowski M, Stepien H, Kunert-Radek J et al. Effect of somatostatin on the proliferation of mouse spleen lymphocytes in vitro. Biochem Biophys Res Commun 1985; 129:52-53.
19. Van Hagen PM, Krenning EP, Kwekkeboom DJ et al. Somatostatin and the immune and haemopoetic system: A review. Eur J Clin Invest 1994; 24:91-99.

20. Stanisz AM, Befus D, Bienebstock J. Differential effects of vasoactive intestinal peptide, substance P, and somatostatin on immunoglobulin synthesis and proliferation by lymphocytes from Payer's patches, mesentheric lymph nodes and spleen. J Immunol 1986; 136:152-156.
21. Cardoso A, El Ghamrawy C, Gautron JP et al. Somatostatin increases mitogen-induced IL-2 secretion and proliferation of human Jurkat T cells via sst3 receptor isotype. J Cell Biochem 1998; 68:62-73.
22. Valatas V, Kolios G, Notas G et al. Octreotide inhibits inflammatory mediators production in rat Kupffer cells. Eur J Endocrinol 2003; 148(suppl. 1):419, (abstr).
23. Komorowski J, Jankiewicz-Wika J, Stepien H. Somatostatin (SOM) and octreotide (OCT) inhibit the secretion of interleukin-8 (IL-8) from human peripheral blood monuclear cells (PBMC) in vitro. Horm Metab Res 2000; 32:337-338.
24. Komorowski J, Jankiewicz-Wika J, Stepien T et al. Octreotide inhibits the secretion of IL-12 from monuclear cells in human peripheral blood (PBMCs)in vitro. Horm Metab Res 2001; 33:689-690.
25. Elliot DE, Weinstock JV. Granulomas in murine schisostomiasis mansoni have a somatostatin immunoregulatory circuit. Metabolism 1996; 45(suppl 1):88-90.
26. Karalis K, Mastorakos G, Chrousos GP et al. Somatostatin analogues suppress the inflammatory reaction in vivo. J Clin Invest 1994; 93:2000-2006.
27. Bjorling DE, Saban MR, Saban R. Effect of octreotide, a somatostatin analogue, on release of inflammatory mediators from isolated guinea pig bladder. J Urol 1997; 158:258-264.
28. Kurnatowska I, Pawlikowski M. Effect of somatostatin analog octreotide on the adjuvant arthritis in rat. Neuroendocrinol Lett 2000; 21:121-126.
29. Kurnatowska I, Pawlikowski M. Anti-inflammatory effects of somatostatin analogs on zymosan-induced earlobe inflamamtion in mice: Comparison with dexamethasone and ketoprofen. Neuroimmunomodulation 2001; 9:119-124.
30. Barrie R, Woltering EA, Hajarizdeh H et al. Inhibition of angiogenesis by somatostatin-like compounds is structurally-dependent. J Surg Res 1993; 55:446-450.
31. Danesi R, Del Tacca M. Effects of octreotide on angiogenesis. In: Scarpignato C, ed. Octreotide: From Basic Science to Clinical Medicine, Vol. 10. Basel: Karger: Prog Basic Clin Pharmacol, 1996:134-245.
32. Pawlikowski M, Kunert-Radek J, Grochal M et al. The effect of somatostatin analog octreotide on diethylstilbestrol -induced prolactin secretion, cell proliferation and vascular changes in the rat anterior pituitary gland. Histol Histopathol 1997; 12:991-994.
33. Lawnicka H, Stepien H, Wyczolkowska J et al. Effect of somatostatin and octreotide on proliferation and vascular endothelial growth factor secretion from murine endothelial cell line (HeCa10) culture. Biochem Biophys Res Commun 2000; 268:567-571.
34. Adams RL, Adams IP, Lindow SW et al. Inhibition of endothelial cell proliferation by the somatostatin analogue SOM 230. Clin Endocrinol 2004; 61:431-436.
35. Pawlikowski M, Kunert-Radek J, Stepien H. Somatostatin inhibits the mitogenic effect of thyroliberin. Experientia 1978; 34:271-272.
36. Pawlikowski M, Kunert-Radek J, Stepien H. Somatostatin- an antiproliferative hormone? In: Dohler KD, Pawlikowski M, eds. Progress in Neuropeptide Research. Basel: Birhauser Verlag, 1989:3-12.
37. Mascardo RN, Sherline P. Somatostatin inhibits rapid centrosomal separation and cell proliferation induced by epidermal growth factor. Endocrinology 1982; 111:1394-1396.
38. Liebow C, Reilly M, Serrano M et al. Somatostatin analogues inhibit growth of pancreatic cancer by stimulating tyrosine phosphatase. Proc Natl Acad Sci USA 1989; 86:2003-2007.
39. Lachowicz-Ochedalska A, Rebas E, Kunert-Radek J et al. Effects of somatostatin and its anlogues on tyrosine kinase activity in rodent tumors. Biol Signals Recept 2000; 9:255-259.
40. Yoshitomi H, Fuji Y, Miyazaki M et al. Involvement of MAP kinase and c-fos signaling in the inhibition of cell growth by somatostatin. Am J Physiol Endocrinol Metab 1997; 272:E769-E774.
41. Gao S, Yu BP, Li Y et al. Antiproliferative effect of octreotide on gastric cancer cells mediated by inhibition of Akt/PKB and telomerase. World J Gastroenterol 2003; 9:2362-2365.
42. Srikant CB. Cell cycle dependent induction of apoptosis by somatostatin analogs by somatostatin analog SMS 201-995 in T-20 mouse pituitary cells. Biochem Biophys Res Commun 1995; 209:400-406.
43. Sharma K, Srikant CB. Induction of wild type p53,Bax, and acid endonuclease during somatostatin-signaled apoptosis in MCF-7 human breast cancer cells. Int J Cancer 1998; 76:259-266.
44. Gruszka A, Kunert-Radek J, Pawlikowski M. The effect of octreotide and bromocriptine on expression of a pro-apoptotic Bax protein in rat prolactinoma. Folia Histochem Cyytobiol 2004; 42:35-39.

CHAPTER 2

Somatostatin Receptors:
Distribution in Normal Tissues and Transduction Mechanisms

Gabriela Melen-Mucha* and Slawomir Mucha

Abstract

Somatostatin (SST) exerts diverse biological effects acting through specific membrane receptors belonging to the family of G protein coupled receptors. So far five SST receptor subtypes (sst_{1-5}) have been cloned and characterized. The genes for these receptor subtypes are localized on different chromosomes and are intronless, with one exception concerning sst_2, which exists in two forms (sst_{2A}, sst_{2B}). Native SST existing mainly in two molecular forms (14 or 28 amino acid peptide) binds to all five sst subtypes with high nanomolar affinity, however some minor differences exist. The commercially available SST analogs (octreotide and lanreotide) bind only with high affinity to sst_2 and sst_5. Based on the binding studies with different SST analogs the family of SST receptors was divided into two subclasses. Subtype 1 and 4 of sst create one subclass and react poorly with these analogs, whereas $sst_{2,3}$ and sst_5 bind these analogs with high affinity comparable to that of native SST.

Structurally, SST receptors consist of seven transmembrane glycoprotein domains connected by short loops (located both extra or intracellularly), one N-terminal extracellular domain with multiple sites for N-linked glycosylation and one C-terminal intracellular domain with motifs for protein phosphorylation.

SST receptor subtypes are variably expressed in distinct tissues both normal and neoplastic. In the brain, pituitary and pancreatic islets all five sst subtypes are present whereas in other peripheral tissues only some of the subtypes are expressed. It should be emphasized that the characteristic pattern of sst subtypes distribution found in the particular organs depends on many factors such as applied methodology, factors regulating sst expression, subtype selectivity, tissue-specificity and species-specificity.

The major intracellular signaling pathways coupled to SST receptors comprise inhibition of adenylyl cyclase leading to the reduction of cAMP levels, activation of phosphotyrosine phosphatases (PTP), modulation of mitogen-activated protein kinases (MAPK) and phospholipase C (PLC), which seem to be universal mediators for all five sst subtypes. Other second messengers including Ca^{2+} channels, K^+ channels, phospolipase A2 (PLA2), Na^+/H^+ exchanger are coupled more selectively with only one or two particular subtypes of sst. Usually, a particular effect of SST and its analogs is mediated through more than one sst subtype (for example GH secretion—sst_2, sst_5). Although, our knowledge concerning SST receptors increases every year, the complete understanding of sst physiology is far to be true. Tachyphylaxis, agonist-induced internalization of sst subtypes and the new fundamental insights concerning

*Corresponding Author: Gabriela Melen-Mucha—Department of Immunoendocrinology Medical University of Lodz, Sterlinga 3, 91-425 Lodz, Poland. Email: g.m-mucha@wp.pl

Somatostatin Analogs in Diagnostics and Therapy, edited by Marek Pawlikowski.
©2007 Landes Bioscience.

homo- and heterodimerization of sst much more complicate sst physiology, but on the other hand create the basis for development of new SST analogs and chimeric molecules to overcome disadvantages of currently available SST analogs.

General Information

Somatostatin (SST) was originally isolated from ovine hypothalami as a growth hormone (GH)-release inhibiting factor in 1973,[1] and was shown to be a cyclic 14 amino acid peptide. The development of antisera against SST by Arimura et al,[2] two years later revealed that SSTs were widely distributed in the body, mainly in neurons of the central and peripheral nervous system and endocrine cells of the pancreas and gastrointestinal tract, which is the main source of SST in the organism.[3] Subsequent studies revealed that SST existed in 2 main forms: 14 amino acid (SST-14) and its N-terminally extended 28 amino acid form (SST-28).[4] Both forms are generated by tissues-specific proteolytic processing of prosomatostatin and are present at various concentrations in different tissues.[5] Moreover, various SST-related peptides were found including cortistatins (CST) first discovered in 1996,[6] which are produced mainly in the brain and immune cells and have similar properties with SST. SSTs are unique in their broad inhibitory effect on both endocrine (inhibition of the secretion of GH, TSH, insulin, glucagon, gasrin, cholecystokinin, vasoactive intestinal hormone, ghrelin) and exocrine secretion (inhibition of the secretion of gastric acid, pancreatic enzymes, intestinal fluid). Moreover, these peptides inhibit various functions of the gastrointestinal tract such as motility, transport, blood flow. Somatostatins influence also the growth processes of normal and neoplastic tissues via inhibition of proliferation and induction of apoptosis, inhibition of angiogenesis and modulation of the immune system.[7-13] Moreover, SSTs change the electrical potential of neurons as other neurotransmitters and modulate several functions of the central and peripheral nervous systems (motor, sensor, behavioral, cognitive), for what they seem to be implicated in a variety of neurological diseases.[14] It was obvious that SSTs influence various processes acting via different ways, as a true hormone, paracrine or autocrine factor and/or neurotransmitter.

History of Somatostatin Receptors

At the beginning it was difficult to conceptualize where the specificity of SST actions laid. On the basis of several biological observations between 1981 and 1985, it was postulated that more than one type of SST receptor might exist to account for the diversity of SST action. One of the first findings speaking in favor of such presumption was the paper showing that both SSTs (14 and 28) have common effects but display different potencies; e.g., SST-28 is more specific inhibitor of pancreatic enzyme secretion, whereas SST-14 is more selective for gastric acid secretion.[15] Moreover, the pharmacological studies and the differential binding properties of both SSTs suggest the existence of at least 2 populations of SST receptors.[16,17] Finally, the synthesis of the first metabolically stable and highly potent SST analog was approved for clinical use, (SMS 201-995, octreotide) in 1982,[18] further permitting identification of two subtypes of SST receptors: one which bounds this compound and another which does not.[17]

In the early 1980s it was well documented that various actions of SSTs are mediated through high affinity membrane receptors which were studied at that time with various methods such as binding analysis using whole cell or plasma membrane fractions, in vivo and in vitro autoradiography, covalent crosslinking and purification of the solubilized SST receptors.[16,17,19-21] With the use of these methods, SST receptors were shown to occur in varying densities in the brain, gut, pituitary, pancreas, adrenals, thyroid and other organs. Moreover, SST receptors were found on several tumor line cells, such as GH_3, AtT-20, Mia PaCa$_2$.[22,23] Photoaffinity labeling and purification studies provide evidence for the existence of at least 3 different forms of SST receptors.[24] Thus, when the efforts to clone the SST receptors began, there was an evidence for 3 different subclasses of SST receptors.

Moreover, in the early 1980s four major intracellular signaling pathways coupled to SST receptors were identified: the inhibition of adenyly cyclase and a fall in intracellular cAMP levels; inhibition of Ca^{2+} channels, modulation of K^+ channels, and stimulation of protein tyrosine phosphatase.[25-27] Almost all of these intracellular mediators are linked and also generally mediated by guanine nucleotide-binding proteins (G proteins).

Molecular Cloning of SST Receptors

In 1992 Yamada et al. cloned the first two SST receptors (SSTR1 and 2), theirs human and mouse homologs.[28] Their strategy was based on reverse transcriptase-polymerase chain reaction (RT-PCR) of highly conserved amino acid sequences in the third and sixth transmembrane segments of G protein-coupled receptors. The specific primers corresponding to these conserved sequences were described by Libert et al.[29] Briefly, RNA was isolated from human pancreatic islet and was reverse-transcribed by using oligo(dT). Then sequences related to G protein coupled receptors were amplified by using set of primers described by Libert et al. The PCR product was labeled with ^{32}P and used to screen a human and mouse genomic libraries what leads to the isolation of the genes encoding the two first SST receptors.[28] Following this discovery, the cloning and sequencing of all five human (h) SST receptors subtypes (hSSTR1,2,3,4,5) (hSSTR3;[30] hSSTR4,[31] hSSTR5,[32]) and their rat homologs (rSSTR3,[33] rSSTR4,[34] rSSTR5,[35]) were reported within 2 years. Few years later they were classified and code-named as sst_{1-5}.[36]

In humans all sst subtypes are encoded by five nonallelic genes located on chromosomes 14 (sst_1), 17 (sst_2), 22 (sst_3), 20 (sst_4) and 16 (sst_5), respectively.[37] The genes encoding $sst_{1,3,4,5}$ are intronless, whereas the gene encoding sst_2 produces the full-length variant sst_{2A} (369 amino acid) and spliced (truncated) variant sst_{2B} (356 amino acids) which differ only in the length of the intracytoplasmic C-tail.[38] The receptor proteins encoded by sst subtype genes are highly conserved in size and structure and display 7 transmembrane domains (TMD) typical of G protein coupled receptors and one N-terminal extracellular and one C-terminal intracellular domain. The SST receptor proteins range in size from 356 to 418 amino acid residues and show the greatest similarity in TMD region (50-70% sequence identity) and the smallest in C-terminal domain. The highest sequence identity of TMD occurs between sst_1 and sst_4 (70%), whereas the smallest between sst_1 an sst_2 (58%). All 5 sst subtypes like other G protein coupled receptors have single or multiple sites for N-linked glycosylation within the extracellular domain and extracellular loop (e.g.: 4 sites for sst_2 and 1 sites for sst_4) and multiple motifs for protein phosphorylation within the intracellular loops and C-terminal domain.[39] Glycosylation increases dramatically the size of receptors as for example in the case of sst_2 from 38 kDa to 85 kDa,[40] whereas phosphorylation may be involved in the receptor regulation.

Pharmacological Characteristics of sst Receptor Proteins

The pharmacological characteristics of sst receptor proteins were studied after introducing the cloned genes into cell lines, such as Chinese hamster ovary (CHO) or kidney cell lines (COS). The pharmacological properties of sst receptor proteins such as ligand binding, G protein coupling and signal transduction pathway in these cell lines were studied in several laboratories. All 5 sst subtypes bind SST-14 and 28 with high nanomolar affinity, though $sst_{1,2,3,4}$ bind SST-14 with 2-3.6 fold higher affinity than SST-28. By contrast sst_5 exhibits 13 fold higher affinity for SST-28 than SST-14, thus SST-28 can be considered as more selective for sst_5, whereas SST-14 is weakly selective what was reported by Patel and Srikant,[41] in CHO-K1 cells. Moreover, the author showed that all 5 sst subtypes exhibited differing selectivities for synthetic SST analogs. Both hexapeptide (MK678) and octapeptide analogs (octreotide, lanreotide, vapreotide) bind to only 3 of the 5 sst subtypes, namely $sst_{2,3,5}$. The striking low affinity of these synthetic analogs for sst_1 and sst_4 was the basis to divide sst family into 2 subclasses: sst_1 and sst_4 create one subclass, whereas $sst_{2,3}$ and sst_5 create another subclass, what corresponds to the previously described SST receptor subtypes 2 and 1, respectively.[42] Furthermore, the affinities of synthetic analogs are comparable to that of SST-14 indicating that they are not selective for a specific subtype, nor more potent than the endogenous ligand.[41]

However, one of the initial report suggested, that a number of octapeptide SST analogs exhibited higher binding affinities (1-2 pM) than endogenous SST and up to 30000 fold binding selectivity for several of the five sst subtypes.[43] These conclusions were based on the binding comparisons of SST receptor subtypes coming from different animal species, expressed in different cell lines and studied with different radioligands, and was not confirmed by other authors.[41,44]

Tissue Distribution of sst Receptor Subtypes

Before cloning of sst receptor subtypes, the detection of sst receptors was performed by ligand binding to the membrane homogenates or tissue slices to detect combined SST receptor binding activity. One of the first studies characterizing sst receptors was shown by Schonbruun and Tashjian.[45] The authors showed the functional characteristics of SST receptor present in the GH4C1 rat pituitary tumor cells with the use of radiolabeled SST ([125I-Tyr1]somatostatin) and conclude that the receptor is necessary for the biological action of SST such as inhibition of prolactin and growth hormone secretion from these cells.

The cloning of 5 sst receptor genes made subtype-specific probes available which allows the investigation of sst receptor subtypes expression at the mRNA level using in situ hybridization (ISH), RNAse protection or reverse transcriptase polymerase chain reaction (RT-PCR) assay.[46,47] While the use of in vitro sst receptor autoradiography, ISH or RT-PCR provided significant information regarding heterogeneity of sst receptor subtypes expressed in normal and tumoral tissues, the precise cellular localization of sst receptor subtypes was difficult to established with these methods. The development of subtype specific polyclonal antibodies in 1995 against sst_1,[48] sst_{2A}[49] and subsequently to other subtypes permitted a precise characterization of the interaction between sst subtype and G proteins and examination of subcellular (membrane or cytoplasm) and intratissular (fibroblast, lymphoid, endothelial or tissue specific cells) distribution of sst receptor subtypes.[48,49] These antibodies were raised against amino acid sequences of sst_1 and sst_{2A} located in the C-terminal region of the receptors. These peptide sequences are conserved in the rat, mouse and human forms. At present, the whole panel of specific antisera against sst subtypes is commercially available and allows us to detect all sst receptor subtypes in normal and tumor tissues,[50] in different assays such as: immunoblot, immunohistochemistry and immunoprecipitation.

Although generally, the mRNA expression of particular substance is assumed to correspond to its protein expression, we should keep in mind that different posttranscriptional processes exist and may change this correlation. This is one of the reasons that explains why such many controversial results exist concerning the expression of sst receptor subtypes detected via different methodological approaches. Moreover, another reason will be the distinct sensitivity and specificity of various methods.

Several studies indicate that the distribution of sst receptor subtypes is very ubiquitous similarly to SST expression. Almost all subtypes of sst receptor are present simultaneously in several tissues and organs such as brain, pituitary, pancreas and gastrointestinal tract in rodents and humans.[28,46,51,52] Whereas in other peripheral tissues the sst receptor subtypes are expressed more selectively, for example mRNA of sst_3 is predominantly expressed in the muscle, sst_4 in the lung.[46]

Distribution of sst Receptor Subtype mRNAs in the Central Nervous System

To determine the expression pattern of all 5 sst subtype mRNAs in different brain regions various methodological approaches were used in the studies.[46,52] Sst_1 and sst_2 mRNAs were widely distributed throughout the central nervous system both in rodents and humans with the highest level in the cortex, amygdala, hypothalamus and hippocampus [46,51-53] However, some species differences were revealed, e.g., high levels of sst_1 mRNA expression in the amygdala in rodents and no expression in these nuclei in humans.[53] Some sex-dependent differences were also noted such as higher expression of sst_1 mRNA in male than female rats in the arcuate nucleus and of sst_2 mRNA in the anterior pituitary, what seems to be involved in the sexually

Table 1. Distribution of sst subtype mRNAs in rodent and human brain[a]

Tissue	sst$_1$	sst$_2$	sst$_3$	sst$_4$	sst$_5$
	\multicolumn{5}{c}{sst Subtypes}				
Amygdala	>++++, -[h]	++,+[h]	+++	+	++,-[h]
Cerebellum	<+	<+, +[h]	>++++, +[h]	-, ++[h]	-, +[h]
Cortex	++++, +[h]	++++, +[h]	++, +[h]	++, -[h]	++, -[h]
Hippocampus	++, +[h]	++, +[h]	++, +[h]	++, +[h]	++, -[h]
Hypothalamus	++, +[h]	++	+	<+	>++++, -[h]
Nucleus accumbens	+	+	<+	+	-
Olfactory bulb	++	++	++	++	++
Preoptic area	+	+	+	+	>++++
Striatum	<+	+	++	+	++
Thalamus	++	+	+	<+	-
Pituitary	++	++, +[h]	+, +[h]	<+,	++, +[h]

[a] If no subscript is used the results are taken from rodent studies. Key: h, human; >++++ to <+, strong to barely detectable signals; -, not detectable. Summary of the data from references 46, 52, 53 and 57.

dimorphic control of some neuroendocrine functions, e.g., pulsatile GH secretion.[54] It should be emphasized that sst$_1$ is colocalized with SST in nerve fibers in hypothalamic nuclei, basal ganglia and retina and in these structures sst$_1$ appears to act as an inhibitory autoreceptor for SST release.[55,56]

The level of sst$_3$ mRNA expression reported by different authors vary quite markedly between different studies, what was discussed in the review by Selmer et al.[52] However consensus exist about the high abundance of sst$_3$ mRNA in the cerebellum and moderate expression in cerebral cortex and hippocampus in rat and human.[46,52,53]

Sst$_4$ is less distributed in the brain than other sst receptor subtypes.[52] The moderate level of sst$_4$ mRNA in rats was shown in the hippocampus, cerebral cortex and olfactory bulb,[52] whereas in human intense expression of sst$_4$ was found in cerebellum, intermediate in hippocampus and no expression in the cortex.[53] In the rat hippocampal and cortical neurons sst$_3$ and sst$_4$ colocalize.[57]

Few studies are available about the distribution of sst$_5$ mRNA. In the rat the highest mRNA content was observed in the hypothalamus and the preoptic area, whereas in other region of rat brain mostly an intermediate level of expression was observed,[52] whereas in human sst$_5$ mRNA expression appears to be very low and restricted to the cerebellum.[53]

The presence of mRNAs for all sst receptor subtypes was demonstrated in the rat pituitary,[46,58] in turn in human pituitary only 3 subtypes of sst (sst$_{2,3,5}$) were present via in situ hybridization.[53]

The published findings are summarized in Table 1.

Distribution of sst Receptor Subtype Proteins in the Central Nervous System

The antisera for sst receptor subtypes produced at the beginning of their development raised against the C-terminal fragment of SST receptors and gave sometimes surprising results. For example antisera against C-terminal part of sst$_1$ revealed very limited distribution of sst$_1$ protein in the rat brain (detected only in the hypothalamus and median eminence),[55] in comparison to the wide distribution of sst$_1$ mRNA.[46] In contrast to this study a more widespread distribution of sst$_1$ protein was identified with antisera to the N-terminal deglycosylated part of sst$_1$.[59] The presence of sst$_1$ was shown throughout the rat brain including the cerebral cortex, hippocampus, basal ganglia and thalamic nuclei. However, it is not clear whether the unglycosylated receptor would be the functional receptor. The development of specific antibodies against all human sst receptor subtypes was a great progress in this field. These antisera were useful tools for mapping the distribution of the individual sst receptor proteins in human tissues.[60] It is worth underlining that the immunohistochemical localization of sst receptors permits investigation of the subcellular distribution of these receptor. For example, electron microscopic evaluation of immunoreactive sst_{2A} in the rat brain revealed that this receptor was located mostly intracellular and only small proportion was associated with the plasma membrane.[61] In the human brain the immunohistochemical localization of sst_{2A} (cerebral cortex, hippocampus) correlates well with the the in situ hybridization studies.[53,62]

In general, most studies based on immunohistochemical localization of sst receptor revealed similar pattern of sst receptor distribution to previously shown through mRNA analysis, what is reviewed by Selmer et al.[52] However, some exceptions exist. For example, in contrast to previously published in situ hybridization analysis of sst_5, in which almost no expression of sst_5 in adult rat brain could be detected, immunohistochemical localization revealed labeling in the basal forebrain, thalamus, cerebral cortex, hippocampus, hypothalamus, amygdala.[63] These discrepancies seem to depend on too low sensitivity of in situ hybridization, because more sensitive RNAse protection experiments did detected sst_5 mRNA in agreement with immunohistochemistry.[52,63]

Distribution of sst Receptor Subtypes in the Periphery

In the periphery the distribution of sst receptor subtypes was also examined at mRNA and protein levels. The studies showed that SST receptors were widely distributed in the gastrointestinal tract, all subtypes are simultaneously expressed in the stomach, jejunum and pancreas (with each gut layer expressing multiple sst receptor subtypes).[64] In rat ocular tissues, all 5 sst receptors were expressed to various extents but, predominantly sst_2 was present in the retina and sst_4 in the iris/ciliary body.[65]

The development of selective antibodies against the human sst receptor subtypes was the great progress in the localization procedures of sst subtypes in human tissues.[60] However, the number of studies concerning systemic immunohistochemical (IHC) distribution of all sst receptor subtypes in human tissues is very limited. In the very recent paper by Taniyama at al[66] the IHC distribution of all subtypes of sst receptors in various human peripheral tissues was shown. In the examined tissues various sst receptor subtypes were detected not only in parenchymal cells but also in stromal cells such as lymphocytes, fibroblasts, and endothelial cells. Among tissues in which the presence of sst receptors has not been previously reported, parotid (sst_{2B} and sst_5), bronchial ($sst_{1,2B,3,4,5}$) and parathyroid gland ($sst_{1,3,4}$) demonstrated immunoreactivity for various subtypes of sst. Moreover, stomach, duodenum and pancreatic islets expressed simultaneously all subtypes of sst receptors, whereas nonislet cells of pancreas are associated with all subtypes except of sst_{2A} and $_{2B}$. Small intestine and colon expressed all sst except sst_4, similar kidney (except sst_3) and liver (except $sst_{3,4}$). In other peripheral tissues a distinct pattern of sst subtypes was observed in the studies and is presented in Table 2.

G Protein Coupling and Signal Transduction of sst Receptors

The different biological actions of SST are initiated by interaction with specific plasma membrane receptors (sst_{1-5}). All 5 sst subtypes bind not only their natural ligands (SST-14 and SST-28) but also cortistatin with high affinity[41,69] and display more selective binding profile for the synthetic SST analogs.[41] Ligand binding induces G protein (guanine triphosphate-binding protein) activation what leads to the activation of several intracellular signaling pathways.

Using various methods such as mutational analysis, transfection studies, immunoprecipitation and affinity purification techniques, a number of G proteins associated with different subtypes of sst have been identified including pertussis toxin (PTX)-sensitive G proteins: $G_{\alpha i1}$, $G_{\alpha i2}$, $G_{\alpha i3}$, as well as PTX-insensitive G proteins: $G_{\alpha q}$, $G_{\alpha 14}$, $G_{\alpha 16}$.[37,49,70-72] What type of G

Table 2. Somatostatin receptor distribution in the peripheral organs of rodent and human[a]

Tissue	sst$_1$	sst$_{2A}$	sst$_{2B}$	sst$_3$	sst$_4$	sst$_5$
Adrenal	+	+	nd	-	-	+
Bronchial gland	+[h]	-[h]	+[h]	+[h]	+[h]	+[h]
Heart	+	-	nd	+	+	-
Kidney	+, +[h]	+, +[h]	+[h]	+,-[h]	+,+[h]	+,+[h]
Lung	+, +[h]	-	nd	+	+, +[h]	-
Muscle	+	-	nd	+	-	+
Ocular tissue	+	+[retina]	nd	+	+[iris]	+
Ovary	+	-	nd	-	-	-
Parathyroid gland	+[h]	-[h]	-[h]	+[h]	+[h]	+[h]
Parotid gland	-[h]	-[h]	+[h]	-[h]	-[h]	+[h]
Placenta	+[h]	-	nd	-	+[h]	+[h]
Prostate	-	-	nd	+	-	-
Spleen	+	+	nd	+	+	+
Testes	+	+	nd	+	-	+
Thyroid	-[h]	-[h]	+[h]	+[h]	+[h]	+[h]
Gastrointestinal tract						
Liver	+,+[h]	+[h]	+[h]	+,-[h]	+,-[h]	+,+[h]
Pancreas	+, +[h]	+, +[h]	+[h]	+,+[h]	+, +[h]	+,+[h]
Stomach	+, +[h]	+,+[h]	+[h]	+,+[h]	+,+[h]	+,+[h]
Small intestine	+, +[h]	+, +[h]	+[h]	+,+[h]	+,-[h]	+,+[h]
Colon	+, +[h]	+, +[h]	nd	+,+[h]	-, -[h]	+, +[h]

[a] If no subscript is used the results are taken from rodent studies. Key: h, human; +, present; -, absent; nd, not determined. The data have been taken from references 28, 46, 52 and 64-68.

protein is involved in the signal transduction of particular subtype of SST receptors depends both on the type of G protein expressed in the examined cells and on the subtype and the level of receptor expression with low affinity receptor/G protein interactions becoming of functional significance at higher receptor densities.[73] Therefore it is not surprising that the use of different cell lines expressing for example high level of transfected sst$_1$ receptors have led to conflicting conclusions regarding the specificity of sst$_1$/G protein coupling. In some cells sst$_1$ is unable to couple to G proteins or inhibit adenylyl cyclase, whereas in others is able.[74,75]

All 5 subtypes of sst are functionally coupled to inhibition of adenylyl cyclase via PTX-sensitive G proteins, what leads to the decrease of cyclic AMP levels,[37,75-77] and may affect various downstream elements such as protein kinase A and subsequent cAMP-response-element-binding protein. Another common signaling pathway activated via all 5 subtypes of sst is a modulation of mitogen-activated protein kinases (MAPK), however the effects vary according to the receptor subtype involved.[37] Sst$_1$ and sst$_4$ activate the MAPK pathway specifically ERK1/2 in transfected CKO-K1 cells. Additionally, sst$_4$-induced activation of MAPK cascade leads to activation of phospholipase A2 (PLA2) and STAT3.[78] On the other hand, sst$_5$ inhibits MAPK pathway via guanylyl cyclase pathway in CHO cells what seems to be implicated in the antiproliferative action of sst$_5$.[79] Activation of phosphotyrosine phosphatases (PTP) through G protein-dependent mechanisms creates another common signaling pathway of sst receptor subtypes. One of the first reports concerning stimulatory effect of SST on PTP activation was published in 1982.[80] Thereafter, numerous studies indicated

Table 3. Sst subtypes signaling pathways

Subtype	sst_1	sst_2	sst_3	sst_4	sst_5
Gene localization	14q13	17q24	22q13.1	20p11.2	16p13.13
Messengers	cAMP↓	cAMP↓	cAMP↓	cAMP↓	cAMP↓
	PTP↑	PTP↑	PTP↑	PTP↑↓	PTP↑
	MAPK↑	MAPK↑↓	MAPK↑↓	MAPK↑	MAPK↓
	PLC/IP3↑	PLC/IP3↑	PLC/IP3↑	PLC/IP3↑	PLC/IP3↑↓
		K^+chan.↑	K^+chan.↑	K^+chan.↑	K^+chan.↑
	Ca^{2+}chan.↓	Ca^{2+}chan.↓			
	Na^+/H^+ exchan.		p53↑, bax↑	PLA2↑	cGMP↓

activation of various types of PTP (SHP-1, SHP-2 and others) via various subtypes of sst. The involvement of various subtypes of sst in PTP activation was shown by Patel in transfected 3T3 and CHO cells ($sst_{1,2,3,4}$),[81] and by other authors.[82,83] The activation of PTP mediate predominantly antiproliferative action of these receptors via inhibition of mitogenic signaling, e.g., SST-activated PTP induce dephosphorylation of receptor thyrosine kinases of various growth factor,[23,27] what counteract the mitogenic signal of the latter or via inhibition of extracellular signal regulated kinase (ERK) and induction of cyclin-dependent kinase inhibitor p27[kip1] leading to cell-cycle arrest.[83]

Moreover, all sst subtypes are coupled to various phospholipase C (PLC) isoforms.[84] SST receptor activation stimulates PLC and increases intracellular levels of inositol triphosphate (IP3) and induces Ca^{2+} mobilization.[72,84,85] Certain subtypes of sst are also coupled to K^+ channels,[86,87] voltage -dependent Ca^{2+} channels,[88] Na^+/H^+ exchanger,[89] and PLA2.[37,90]

It has been emphasized that various actions of SST are mediated through various sst subtypes implicating various intracellular signaling pathways. For example, SST inhibits secretion of several hormones and enzymes mainly via inhibition of intracellular cAMP and Ca^{2+} levels and by the influence of receptor-linked exocytosis.[37,91] This hormone inhibits the growth of several normal and neoplastic tissues not only through the inhibition of cell proliferation (through all sst subtypes) but also via induction of apoptosis through sst_2 and sst_3 subtypes. The induction of apoptosis can be cell cycle dependent as in AtT-20 mouse pituitary cell,[92] or independent and signaled mainly via sst_3 subtype through the wild type of p53 and bax[11] (see Table 3).

Other Aspects of sst Receptor Subtype Physiology

Although our understanding of sst subtype pathophysiology increases every year, we should emphasize that most of the above-mentioned findings concern the sst receptor subtypes expressed by transfection into various cell lines but not endogenous receptors. Studies of endogenous SST receptors ensure that the properties of the receptor under study are being characterized in their native environment in the presence of all the cellular machinery required for normal function. However such studies are difficult because most SST-responsive tissues and many cell lines express several subtypes of sst. Moreover, endogenous sst subtypes are generally expressed at very low levels and thus biochemical studies are limited by the availability of material for study. Furthermore, the whole panel of specific agonist and antagonist for different sst subtypes was not available for a long time, so it was not possible to determine which receptor is responsible for the particular biological action of SST in the native environment. Although the first agonist peptides (not specific) were developed very early after SST discovery, the structure of the first antagonist peptide was developed only in 1996.[93]

Today almost a complete panel of antagonists seems to exist, however some of them display also agonist-like properties.[94,95]

Thus, in these days our understanding of the pathophysiology of endogenously expressed sst subtypes is far of being completed.

Moreover, agonist-induced internalization or upregulation of sst subtypes further complicate our knowledge about sst physiology.[96] The receptor internalization is common among G protein receptors. Ligand-induced internalization of various sst receptor subtypes depends on exposure time, ligand concentration and other factors. Human $sst_{2,3,4}$ and $_5$ are internalized efficiently and rapidly whereas sst_1 displayed virtually no internalization in CHO-K1 cells.[96] Prolonged agonist treatment led to different up- or down-regulation of sst receptors what seems to be subtype selective (in CHO-K1 cells, SST upregulates mainly $sst_{1,4}$ in contrast to $sst_{3,5}$),[96] and depends on several other factors such as repeated or continuous mode of agonist administration and others.[97] Internalization of sst receptor subtypes may be one of the mechanisms that regulate the number of sst receptor sites on cell surface and seems to be the most important in sst-targeted radio- and chemotherapy.[98] Moreover, the internalization of sst receptors seems to be one of the mechanisms involved in the adaptation or tachyphylaxis observed during continuous exposure to SST or SST analogs. Moreover, the desensitization of responsiveness to SST may be associated with other processes such as receptor phosphorylation, G protein uncoupling and receptor degradation and down-regulation.[98] It should be emphasized that the loss of responsiveness to SST analogs is a very frequent problem in the treatment of gastroenteropancreatic tumors[99] while is almost not observed in acromegalic patients.[100]

The new aspect of sst receptor physiology was revealed by Rocheville et al. in 2000.[101,102] The quoted authors showed that sst receptor subtypes exhibited ligand-induced homo- and hetero-dimerization in transfected CHO-K1 cells, what often changed their binding and other functional properties. For example, sst_5 forms homodimers as well as heterodimers with another subtype of SST receptor—sst_1. Moreover, sst receptors can heterodimerize also with other classes of G protein receptors such as dopamine 2 receptor (D2) (sst_5 and D2)[101] or opioid receptors which are the nearest relatives of sst (sst_{2A} with μ-opioid).[103] These findings were the basis for the development of chimeric somatostatin/dopamine molecule (BIM23A387), which suppresses GH or PRL release from pituitary adenomas more potently than the sst2 and D2 components given separately as well as in combination.[104] The development of various chimeric molecules to enhance some physiological effects of SST and its analogs is now intensively studied.

Molecular Genetic Approaches to Studying sst Receptor Subtype Function

In 1997 Zheng et al[105] described the generation of the first sst receptor gene knock-out, sst_2 deficient mouse (sst_2 KO). These animals are reported to be normal and healthy but they are refractory to GH-induced feedback suppression of GH due to a loss of sst_2 in arcuate GHRH neurons. Nowadays, the majority of the relevant data concerning the selective involvement of different sst subtypes in specific physiological responses to SST came from the studies on KO mice, in which gene encoding for a given sst receptor subtype has been invalidated. These animals are now available for each of the five sst receptor subtypes (for sst_{1-4},[106] and for sst_5,[107]), but the majority of studies concern sst_2 and sst_5. The studies of sst_2 KO mice indicate that this receptor displays some specific central and peripheral actions, such as fine motor control,[108] modulation of learning,[109] and inhibition of gastric acid secretion.[110] Analysis of the sst_5 KO phenotype showed that this receptor mediated inhibition of insulin secretion and glucose homeostasis.[107]

However, the data obtained in KO models should be interpreted with caution because the genetic deletion of for ex. sst_1 triggers an over-expression of sst_2 and vice versa.[111]

Conclusions

Although, the huge number of papers concerning the role of transfected sst receptor subtypes is published every year, our knowledge about pathophysiological significance of endogenous receptor is still obscure. The development of almost whole panel of sst receptor specific agonist and antagonist together with the generation of sst receptor KO mice allows us to evaluate the role of endogenous receptors in their natural environment and to draw the right conclusions concerning their significance in several pathological conditions.

Acknowledgements

This review was supported by a grant no. 502-11-295 from Medical University of Lodz.

References

1. Brazeau P, Vale W, Burgus R et al. Hypothalamic polypeptide that inhibits the secretion of immunoreactive pituitary growth hormone. Science 1973; 179:77-79.
2. Arimura A, Sato H, Dupont A et al. Somatostatin: Abundance of immunoreacitve hormone in the rat stomach and pancreas. Science 1975; 189:10007-9.
3. Patel YC, Wheatley T, Ning C. Multiple forms of immunoreactive somatostatin: Comparison of distribution in neural and nonneural tissues and portal plasma of the rat. Endocrinology 1981; 109:1943-49.
4. Pradayrol L, Jornvall H, Mutt V et al. N-terminally extended somatostatin: The primary structure of Somatostatin-28. FEBS Lett 1980; 109:55-58.
5. Shen LP, Pictet RL, Rutter WJ. Human somatostatin I: Sequence of the cDNA. Proc Natl Acad Sci USA 1982; 79(15):4575-9.
6. de Lecea L, Criado JR, Prospero-Garcia O et al. A cortical neuropeptide with neuronal depressant and sleep-modulating properties. Nature 1996; 381(6579):242-5.
7. Guillemin R, Gerich JE. Somatostatin: Physiological and clinical significance. Annu Rev Med 1976; 27:379-88
8. Yamada T. Gut somatostatin. In: Reichlin S, ed. Somatostatin. Basic and Clinical Status. New York: Plenum Press, 1987:221-228.
9. Pawlikowski M, Kunert-Radek J, Stepien H. Somatostatin inhibits the mitogenic effect of thyroliberin. Experientia 1978; 34:271-271
10. Mascardo RN, Sherline P. Somatostatin inhibits rapid centrosomal separation and cell proliferation induced by epidermal growth factor. Endocrinology 1982; 111(4):1394-6.
11. Sharma K, Patel YC, Srikant CB. Subtype-selective induction of wild-type p53 and apoptosis, but not cell cycle arrest, by human somatostatin receptor 3. Mol Endocrinol 1996; 10(12):1688-96.
12. Woltering EA. Development of targeted somatostatin-based antiangiogenic therapy: A review and future perspectives. Cancer Biother Radiopharm 2003; 18(4):601-9.
13. Ten Bokum AMC, Hofland LJ, van Hagen. Somatostatin and somatostatin receptors in the immune system: A review. Eur Cytokine Net 2000; 11:161-7.
14. Schindler M, Humphrey PP, Emson PC. Somatostatin receptors in the central nervous system. Prog Neurobiol 1996; 50(1):9-47.
15. Reichlin S. Somatostatin. N Engl J Med 1983; 309(25):1495-1501, (1556-63).
16. Srikant CB, Patel YC. Receptor binding of somatostatin-28 is tissue specific. Nature 1981; 294(5838):259-60
17. Reubi JC. Evidence for two somatostatin-14 receptor types in rat brain cortex. Neurosci Lett 1984; 49(3):259-63.
18. Bauer W, Briner U, Doepfner W et al. SMS 201-995: A very potent and selective octapeptide analogue of somatostatin with prolonged action. Life Sci 1982; 31(11):1133-40.
19. Srikant CB, Patel YC. Somatostatin receptors: Identification and characterization in rat brain membranes. Proc Natl Acad Sci USA 1981; 78(6):3930-4.
20. Patel YC, Murthy KK, Escher EE et al. Mechanism of action of somatostatin: An overview of receptor function and studies of the molecular characterization and purification of somatostatin receptor proteins. Metabolism 1990; 39(9 Suppl 2):63-9.
21. Reubi JC, Cortes R, Maurer R et al. Distribution of somatostatin receptors in the human brain: An autoradiographic study. Neuroscience 1986; 18(2):329-46.
22. Thermos K, Reisine T. Somatostatin receptor subtypes in the clonal anterior pituitary cell lines AtT-20 and GH3. Mol Pharmacol 1988; 33(4):370-7.

23. Hierowski MT, Liebow C, du Sapin K et al. Stimulation by somatostatin of dephosphorylation of membrane proteins in pancreatic cancer MIA PaCa-2 cell line. FEBS Lett 1985; 179(2):252-6.
24. Srikant CB, Murthy KK, Escher EE et al. Photoaffinity labeling of the somatostatin receptor: Identification of molecular subtypes. Endocrinology 1992; 130(5):2937-46.
25. Koch BD, Schonbrunn A. The somatostatin receptor is directly coupled to adenylate cyclase in GH4C1 pituitary cell membranes. Endocrinology 1984; 114(5):1784-90.
26. Koch BD, Blalock JB, Schonbrunn A. Characterization of the cyclic AMP-independent actions of somatostatin in GH cells. I. An increase in potassium conductance is responsible for both the hyperpolarization and the decrease in intracellular free calcium produced by somatostatin. J Biol Chem 1988; 263(1):216-25.
27. Liebow C, Reilly C, Serrano M et al. Somatostatin analogues inhibit growth of pancreatic cancer by stimulating tyrosine phosphatase. Proc Natl Acad Sci USA 1989; 86(6):2003-7.
28. Yamada Y, Post SR, Wang K et al. Cloning and functional characterization of a family of human and mouse somatostatin receptors expressed in brain, gastrointestinal tract, and kidney. Proc Natl Acad Sci USA 1992; 89(1):251-5.
29. Libert F, Parmentier M, Lefort A et al. Selective amplification and cloning of four new members of the G protein-coupled receptor family. Science 1989; 244(4904):569-72.
30. Yamada Y, Reisine T, Law SF et al. Somatostatin receptors, an expanding gene family: Cloning and functional characterization of human SSTR3, a protein coupled to adenylyl cyclase. Mol Endocrinol 1992; 6(12):2136-42.
31. Rohrer L, Raulf F, Bruns C et al. Cloning and characterization of a fourth human somatostatin receptor. Proc Natl Acad Sci USA 1993; 90(9):4196-200.
32. Panetta R, Greenwood MT, Warszynska A et al. Molecular cloning, functional characterization, and chromosomal localization of a human somatostatin receptor (somatostatin receptor type 5) with preferential affinity for somatostatin-28. Mol Pharmacol 1994; 45(3):417-27.
33. Meyerhof W, Wulfsen I, Schonrock C et al. Molecular cloning of a somatostatin-28 receptor and comparison of its expression pattern with that of a somatostatin-14 receptor in rat brain. Proc Natl Acad Sci USA 1992; 89(21):10267-71.
34. Bruno JF, Xu Y, Song J et al. Molecular cloning and functional expression of a brain-specific somatostatin receptor. Proc Natl Acad Sci USA 1992; 89(23):11151-5.
35. O'Carroll AM, Lolait SJ, Konig M et al. Molecular cloning and expression of a pituitary somatostatin receptor with preferential affinity for somatostatin-28. Mol Pharmacol 1992; 42(6):939-46.
36. Hoyer D, Bell GI, Berelowitz M et al. Classification and nomenclature of somatostatin receptors. Trends Pharmacol Sci 1995; 16(3):86-8.
37. Patel YC. Somatostatin and its receptor family. Front Neuroendocrinol 1999; 20(3):157-98.
38. Patel YC, Greenwood M, Kent G et al. Multiple gene transcripts of the somatostatin receptor SSTR2: Tissue selective distribution and cAMP regulation. Biochem Biophys Res Commun 1993; 192(1):288-94.
39. Patel YC, Greenwood MT, Panetta R et al. The somatostatin receptor family. Life Sci 1995; 57(13):1249-65, (Review).
40. Eppler CM, Zysk JR, Corbett M et al. Purification of a pituitary receptor for somatostatin. The utility of biotinylated somatostatin analogs. J Biol Chem 1992; 267(22):15603-12.
41. Patel YC, Srikant CB. Subtype selectivity of peptide analogs for all five cloned human somatostatin receptors (hsstr 1-5). Endocrinology 1994; 135(6):2814-7.
42. Raynor K, Reisine T. Analogs of somatostatin selectively label distinct subtypes of somatostatin receptors in rat brain. J Pharmacol Exp Ther 1989; 251(2):510-7.
43. Raynor K, Murphy WA, Coy DH et al. Cloned somatostatin receptors: Identification of subtype-selective peptides and demonstration of high affinity binding of linear peptides. Mol Pharmacol 1993; 43(6):838-44.
44. Bruns C, Weckbecker G, Raulf F et al. Molecular pharmacology of somatostatin-receptor subtypes. Ann NY Acad Sci 1994; 733:138-46.
45. Schonbrunn A, Tashjian Jr H. Characterization of functional receptors for somatostatin in rat pituitary cells in culture. J Biol Chem 1978; 253(18):6473-83.
46. Raulf F, Perez J, Hoyer D et al. Differential expression of five somatostatin receptor subtypes, SSTR1-5, in the CNS and peripheral tissue. Digestion 1994; 55(suppl 3):46-53.
47. Reubi JC, Schaer JC, Waser B et al. Expression and localization of somatostatin receptor SSTR1, SSTR2, and SSTR3 messenger RNAs in primary human tumors using in situ hybridization. Cancer Res 1994; 54(13):3455-9.
48. Gu YZ, Brown PJ, Loose-Mitchell DS et al. Development and use of a receptor antibody to characterize the interaction between somatostatin receptor subtype 1 and G proteins. Mol Pharmacol 1995; 48(6):1004-14.

49. Gu YZ, Schonbrunn A. Coupling specificity between somatostatin receptor sst2A and G proteins: Isolation of the receptor-G protein complex with a receptor antibody. Mol Endocrinol 1997; 11(5):527-37.
50. Pawlikowski M, Pisarek H, Kunert-Radek J et al. Immunohistochemical detection of somatostatin receptor subtypes in "clinically nonfunctioning" pituitary adenomas. Endocr Pathol 2003; 14(3):231-8.
51. Bruno JF, Xu Y, Song J et al. Tissue distribution of somatostatin receptor subtype messenger ribonucleic acid in the rat. Endocrinology 1993; 133(6):2561-7.
52. Selmer I, Schindler M, Allen JP et al. Advances in understanding neuronal somatostatin receptors. Regul Pept 2000; 90(1-3):1-18. (Review).
53. Thoss VS, Perez J, Probst A et al. Expression of five somatostatin receptor mRNAs in the human brain and pituitary. Naunyn Schmiedebergs Arch Pharmacol 1996; 354(4):411-9.
54. Zhang WH, Beaudet A, Tannenbaum GS. Sexually dimorphic expression of sst1 and sst2 somatostatin receptor subtypes in the arcuate nucleus and anterior pituitary of adult rats. J Neuroendocrinol 1999; 11(2):129-36.
55. Helboe L, Stidsen CE, Moller M. Immunohistochemical and cytochemical localization of the somatostatin receptorsubtype sst1 in the somatostatinergic parvocellular neuronal system of the rat hypothalamus. J Neurosci 1998; 18(13):4938-45.
56. Thermos K, Bagnoli P, Epelbaum J et al. The somatostatin sst(1) receptor: An autoreceptor for somatostatin in brain and retina? Pharmacol Ther 2005; 3, [Epub ahead of print].
57. Perez J, Hoyer D. Coexpression of somatostatin SSTR-3 and SSTR-4 receptor messenger RNAs in the rat brain. Neuroscience 1995; 64(1):241-53.
58. Day R, Dong W, Panetta R et al. Expression of mRNA for somatostatin receptor (sstr) types 2 and 5 in individual rat pituitary cells. A double labeling in situ hybridization analysis. Endocrinology 1995; 136(11):5232-5.
59. Hervieu G, Emson PC. The localization of somatostatin receptor 1 (sst1) immunoreactivity in the rat brain using an N-terminal specific antibody. Neuroscience 1998; 85(4):1263-84.
60. Helboe L, Moller M, Norregaard L et al. Development of selective antibodies against the human somatostatin receptor subtypes sst1-sst5. Brain Res Mol Brain Res 1997; 49(1-2):82-8.
61. Dournaud P, Boudin H, Schonbrunn A et al. Interrelationships between somatostatin sst2A receptors and somatostatin-containing axons in rat brain: Evidence for regulation of cell surface receptors by endogenous somatostatin. J Neurosci 1998; 18(3):1056-71.
62. Schindler M, Holloway S, Humphrey PP et al. Localization of the somatostatin sst2(a) receptor in human cerebral cortex, hippocampus and cerebellum. Neuroreport 1998; 9(3):521-5.
63. Stroh T, Kreienkamp HJ, Beaudet A. Immunohistochemical distribution of the somatostatin receptor subtype 5 in the adult rat brain: Predominant expression in the basal forebrain. J Comp Neurol 1999; 412(1):69-82.
64. Krempels K, Hunyady B, O'Carroll AM et al. Distribution of somatostatin receptor messenger RNAs in the rat gastrointestinal tract. Gastroenterology 1997; 112(6):1948-60.
65. Mori M, Aihara M, Shimizu T. Differential expression of somatostatin receptors in the rat eye: SSTR4 is intensely expressed in the iris/ciliary body. Neurosci Lett 1997; 223(3):185-8.
66. Taniyama Y, Suzuki T, Mikami Y et al. Systemic distribution of somatostatin receptor subtypes in human: An immunohistochemical study. Endocr J 2005; 52(5):605-11.
67. Caron P, Buscail L, Beckers A et al. Expression of somatostatin receptor SST4 in human placenta and absence of octreotide effect on human placental growth hormone concentration during pregnancy. J Clin Endocrinol Metab 1997; 82(11):3771-6.
68. Zhu LJ, Krempels K, Bardin CW et al. The localization of messenger ribonucleic acids for somatostatin receptors 1, 2, and 3 in rat testis. Endocrinology 1998; 139(1):350-7.
69. Siehler S, Seuwen K, Hoyer D. [125I]Tyr10-cortistatin14 labels all five somatostatin receptors. Naunyn Schmiedebergs Arch Pharmacol 1998; 357(5):483-9.
70. Luthin DR, Eppler CM, Linden J. Identification and quantification of Gi-type GTP-binding proteins that copurify with a pituitary somatostatin receptor. J Biol Chem 1993; 268(8):5990-6.
71. Ferjoux G, Bousquet C, Cordelier P et al. Signal transduction of somatostatin receptors negatively controlling cell proliferation. J Physiol Paris 2000; 94(3-4):205-10, (Review).
72. Rosskopf D, Schurks M, Manthey I et al. Signal transduction of somatostatin in human B lymphoblasts. Am J Physiol Cell Physiol 2003; 284(1):C179-90.
73. Milligan G. Mechanisms of multifunctional signalling by G protein-linked receptors. Trends Pharmacol Sci 1993; 14(6):239-44, (Review).
74. Rens-Domiano S, Law SF, Yamada Y et al. Pharmacological properties of two cloned somatostatin receptors. Mol Pharmacol 1992; 42(1):28-34.

75. Hershberger RE, Newman BL, Florio T et al. The somatostatin receptors SSTR1 and SSTR2 are coupled to inhibition of adenylyl cyclase in Chinese hamster ovary cells via pertussis toxin-sensitive pathways. Endocrinology 1994; 134(3):1277-85.

76. Patel YC, Greenwood MT, Warszynska A et al. All five cloned human somatostatin receptors (hSSTR1-5) are functionally coupled to adenylyl cyclase. Biochem Biophys Res Commun 1994; 198(2):605-12.

77. Meyerhof W. The elucidation of somatostatin receptor functions: A current view. Rev Physiol Biochem Pharmacol 1998; 133:55-108, (Review).

78. Sellers LA, Feniuk W, Humphrey PP et al. Activated G protein-coupled receptor induces tyrosine phosphorylation of STAT3 and agonist-selective serine phosphorylation via sustained stimulation of mitogen-activated protein kinase. Resultant effects on cell proliferation. J Biol Chem 1999; 274(23):16423-30.

79. Cordelier P, Esteve JP, Bousquet C et al. Characterization of the antiproliferative signal mediated by the somatostatin receptor subtype sst5. Proc Natl Acad Sci USA 1997; 94(17):9343-8.

80. Reyl FJ, Lewin MJ. Intracellular receptor for somatostatin in gastric mucosal cells: Decomposition and reconstitution of somatostatin-stimulated phosphoprotein phosphatases. Proc Natl Acad Sci USA 1982; 79(4):978-82.

81. Patel YC. Molecular pharmacology of somatostatin receptor subtypes. J Endocrinol Invest 1997; 20(6):348-67.

82. Reardon DB, Wood SL, Brautigan DL et al. Activation of a protein tyrosine phosphatase and inactivation of Raf-1 by somatostatin. Biochem J 1996; 314:401-4.

83. Florio T, Arena S, Thellung S et al. The activation of the phosphotyrosine phosphatase eta (r-PTP eta) is responsible for the somatostatin inhibition of PC Cl3 thyroid cell proliferation. Mol Endocrinol 2001; 15(10):1838-52.

84. Akbar M, Okajima F, Tomura H et al. Phospholipase C activation and Ca2+ mobilization by cloned human somatostatin receptor subtypes 1-5, in transfected COS-7 cells. FEBS Lett 1994; 348(2):192-6.

85. Wilkinson GF, Feniuk W, Humphrey PP. Characterization of human recombinant somatostatin sst5 receptors mediating activation of phosphoinositide metabolism. Br J Pharmacol 1997; 121(1):91-6.

86. White RE, Schonbrunn A, Armstrong DL. Somatostatin stimulates Ca(2+)-activated K+ channels through protein dephosphorylation. Nature 1991; 351(6327):570-3.

87. Kreienkamp HJ, Honck HH, Richter D. Coupling of rat somatostatin receptor subtypes to a G-protein gated inwardly rectifying potassium channel (GIRK1). FEBS Lett. 1997; 419(1):92-4.

88. Lewis DL, Weight FF, Luini A. A guanine nucleotide-binding protein mediates the inhibition of voltage-dependent calcium current by somatostatin in a pituitary cell line. Proc Natl Acad Sci USA 1986; 83(23):9035-9.

89. Barber DL, McGuire ME, Ganz MB. Beta-adrenergic and somatostatin receptors regulate Na-H exchange independent of cAMP. J Biol Chem 1989; 264(35):21038-42.

90. Cervia D, Fiorini S, Pavan B et al. Somatostatin (SRIF) modulates distinct signaling pathways in rat pituitary tumor cells; negative coupling of SRIF receptor subtypes 1 and 2 to arachidonic acid release. Naunyn Schmiedebergs Arch Pharmacol 2002; 365(3):200-9.

91. Renstrom E, Ding WG, Bokvist K et al. Neurotransmitter-induced inhibition of exocytosis in insulin-secreting beta cells by activation of calcineurin. Neuron 1996; 17(3):513-22.

92. Srikant CB. Cell cycle dependent induction of apoptosis by somatostatin analog SMS 201-995 in AtT-20 mouse pituitary cells. Biochem Biophys Res Commun 1995; 209(2):400-6.

93. Bass RT, Buckwalter BL, Patel BP et al. Identification and characterization of novel somatostatin antagonists. Mol Pharmacol 1996; 50(4):709-15.

94. Nunn C, Langenegger D, Hurth K et al. Agonist properties of putative small-molecule somatostatin sst2 receptor-selective antagonists. Eur J Pharmacol 2003; 465(3):211-8.

95. van der Hoek J, Hofland LJ, Lamberts SW. Novel subtype specific and universal somatostatin analogues: Clinical potential and pitfalls. Curr Pharm Des 2005; 11(12):1573-92.

96. Hukovic N, Panetta R, Kumar U et al. Agonist-dependent regulation of cloned human somatostatin receptor types 1-5 (hSSTR1-5): Subtype selective internalization or upregulation. Endocrinology 1996; 137(9):4046-9.

97. Froidevaux S, Hintermann E, Torok M et al. Differential regulation of somatostatin receptor type 2 (sst 2) expression in AR4-2J tumor cells implanted into mice during octreotide treatment. Cancer Res 1999; 59(15):3652-7.

98. Hofland LJ, Lamberts SW. The pathophysiological consequences of somatostatin receptor internalization and resistance. Endocr Rev 2003; 24(1):28-47, (Review).

99. Arnold R, Simon B, Wied M. Treatment of neuroendocrine GEP tumours with somatostatin analogues: A review. Digestion 2000; 62(Suppl 1):84-91, (Review).
100. Lamberts SW, van der Lely AJ, de Herder WW et al. Octreotide. N Engl J Med 1996; 334(4):246-54.
101. Rocheville M, Lange DC, Kumar U et al. Receptors for dopamine and somatostatin: Formation of hetero-oligomers with enhanced functional activity. Science 2000; 288(5463):154-7.
102. Rocheville M, Lange DC, Kumar U et al. Subtypes of the somatostatin receptor assemble as functional homo- and heterodimers. J Biol Chem 2000; 275(11):7862-9.
103. Pfeiffer M, Koch T, Schroder H et al. Heterodimerization of somatostatin and opioid receptors cross-modulates phosphorylation, internalization, and desensitization. J Biol Chem 2002; 277(22):19762-72.
104. Saveanu A, Lavaque E, Gunz G et al. Demonstration of enhanced potency of a chimeric somatostatin-dopamine molecule, BIM-23A387, in suppressing growth hormone and prolactin secretion from human pituitary somatotroph adenoma cells. J Clin Endocrinol Metab 2002; 87(12):5545-52.
105. Zheng H, Bailey A, Jiang MH et al. Somatostatin receptor subtype 2 knockout mice are refractory to growth hormone-negative feedback on arcuate neurons. Mol Endocrinol 1997; 11(11):1709-17.
106. Videau C, Hochgeschwender U, Kreienkamp HJ et al. Characterisation of [125I]-Tyr0DTrp8-somatostatin binding in sst1- to sst4- and SRIF-gene-invalidated mouse brain. Naunyn Schmiedebergs Arch Pharmacol 2003; 367(6):562-71.
107. Strowski MZ, Kohler M, Chen HY et al. Somatostatin receptor subtype 5 regulates insulin secretion and glucose homeostasis. Mol Endocrinol 2003; 17(1):93-106.
108. Allen JP, Hathway GJ, Clarke NJ et al. Somatostatin receptor 2 knockout/lacZ knockin mice show impaired motor coordination and reveal sites of somatostatin action within the striatum. Eur J Neurosci 2003; 17(9):1881-95.
109. Dutar P, Vaillend C, Viollet C et al. Spatial learning and synaptic hippocampal plasticity in type 2 somatostatin receptor knock-out mice. Neuroscience 2002; 112(2):455-66
110. Martinez V, Curi AP, Torkian B et al. High basal gastric acid secretion in somatostatin receptor subtype 2 knockout mice. Gastroenterology 1998; 114(6):1125-32.
111. Casini G, Dal Monte M, Petrucci C et al. Altered morphology of rod bipolar cell axonal terminals in the retinas of mice carrying genetic deletion of somatostatin subtype receptor 1 or 2. Eur J Neurosci 2004; 19(1):43-54.

CHAPTER 3

Somatostatin Analogs:
Lessons in Functional Complexity and Receptor Interactions

Michael D. Culler*

Abstract

Somatostatin (SST) was originally isolated from the hypothalamus as the key suppressor GH secretion, but is now known to be widely distributed throughout the body and to be involved in numerous physiological functions. While the varied activity of SST creates opportunities for its use as a therapeutic agent, it also hampers drug development by creating the potential for inducing undesired actions. The discovery of five distinct somatostatin receptor (SSTR) subtypes has provided an opportunity for creating subtype-specific, and, potentially, action-specific SST analogs. Through efforts to determine the functional association between receptor subtype and function, it has become apparent that most SST-responsive target tissues express multiple receptor subtypes, and that it is the interaction of the subtypes that will determine the cellular response. In this regard, we have observed both action-enhancing and antagonistic interactions between different SSTR subtype combinations, and that these effects can be induced in a gradated fashion depending on the degree to which the different subtypes present are activated. These observations imply that the cellular response to SST is dictated by the ratio of receptor subtypes present on the cell at a given moment. In this way, the cellular response to SST can be influenced by the prevailing environmental conditions, physiological status and hormonal milieu, through alteration of the ratio of receptor subtypes expressed. One can also envision the development of pathologies in which the cellular response is inappropriate due to altered ratios of receptor subtype expression. The concept of receptor interaction attains even greater complexity with the demonstration that the SSTR subtypes can also interact with members of other receptor families. These complex interactions provide potential opportunities for creating drugs that target only specific combinations of receptors that are expressed under the appropriate conditions to generate a specific cellular response. Learning to unravel this complex cellular code is key, both for furthering our understanding of cellular based physiology and disease, and in order to create more effective drugs with greater functional specificity.

Introduction

It has now been over thirty years since the identification of somatostatin (SST) as the hypothalamic factor responsible for the suppression of growth hormone (GH) secretion from the pituitary.[1] Since that initial discovery, one of the most notable observations emerging from the research surrounding SST physiology is the diversity of biological functions in which SST

*Michael D. Culler—Biomeasure Incorporated/IPSEN, 27 Maple Street, Milford, Massachusetts 01748, U.S.A. Email: michael.culler@ipsen.com

Somatostatin Analogs in Diagnostics and Therapy, edited by Marek Pawlikowski.
©2007 Landes Bioscience.

plays a role.[2-4] The varied activity of SST creates opportunities for its use as a therapeutic agent; however, it also hampers drug development by creating the potential for inducing undesired actions. This issue raises the intriguing question of how nature achieves functional selectivity with a broadly active compound such as SST. Part of the answer resides in the fact that SST is not truly a hormone in the classical sense of being produced in one tissue, carried through the blood, and inducing a response in another tissue. Rather, SST is more properly characterized as a paracrine or autocrine factor, in that all tissues in which SST receptors are found have a local source of SST production. Even with the originally identified function for SST, suppression of GH secretion, SST is carried only millimeters within the specialized hypophyseal portal capillary system to reach the GH-producing cells of the pituitary. Within the small fluid volume of the hypophyseal capillaries and the extracellular compartments of tissues, a small release of SST achieves a tremendously high concentration at its target receptors. Once the locally produced SST is washed into the general circulation, however, there is an immediate dilution that lowers SST concentration to a level well below that required for receptor activation. In addition, once in the circulation SST is rapidly inactivated by enzymatic degradation and cleared,[5] thus preventing accumulation over time. In this way, the SST produced in one tissue in order to generate a response to a particular physiologic need does not reach and interact with the receptors of another tissue.

These observations provide useful insight into a potential means by which the actions of SST at different sites are kept distinct, but present a problem for development of SST analogs as drugs. Ideally, a SST analog should be delivered directly to the site of desired action. However, since targeted administration is not always feasible with current drug delivery technologies, in practice, SST analogs are normally delivered systemically. This creates the problem of having to produce a high systemic concentration of the drug in order to achieve a sufficiently high concentration of the drug within the target tissue to activate the target receptors. By definition, if the drug concentration is sufficiently high to affect the receptors in the desired target tissue, it is sufficiently high to affect the same or similar receptors in other tissues, and potentially cause dysregulation and undesired side effects. Consequently, the challenge is to produce analogs of SST that can achieve sufficient concentrations to activate the desired target receptors, while not inducing unwanted effects in nontarget tissues.

The problem of selectivity for SST analogs was first considered during optimization of the two SST analogs currently used for treatment of excess GH secretion, Octreotide and Lanreotide. It was known that in addition to suppressing GH, SST also possessed the undesirable effect of suppressing insulin secretion. In early studies it was noted that SST analogs had differing ratios of activity with regard to GH and insulin secretion.[6-9] Using in vivo bioassays, analogs were screened and optimized to yield the two analogs currently used clinically, which are very potent at suppressing GH secretion, but which also exhibit acceptably low insulin-suppressing activity.[10,11]

The reason that different analogs had differing ratios of pituitary and pancreatic activity became apparent in the early 1990s with the elucidation of five distinct SST receptor (SSTR) subtypes (SSTR1-5).[12] It was found that both Octreotide and Lanreotide are highly potent at the SSTR2 subtype, which we now know is the primary SSTR responsible for suppressing GH secretion in rats,[13] but have only moderate activity for the SSTR5 subtype, which we now know to be the primary SSTR responsible for suppressing insulin secretion through direct interaction with the pancreatic β-cells.[14,15] The discovery of multiple SSTR subtypes also suggests another mechanism used by nature to achieve functional specificity for SST by having unique receptors control individual functions. If this concept is correct, it also suggests a potential means of achieving functional specificity for SST-based drugs by producing SST analogs that are selective for a specific receptor subtype, and, thereby, a specific biological function. The challenge then becomes one of elucidating which SSTR subtype controls a specific biological function. As discussed in the following sections, however, both the task of determining the involvement of the SSTR subtypes in particular functions, and the actual physiological organization of the SSTRs, are far more complex than originally anticipated.

Matching SSTR Subtypes with Function—Not an Easy Task

One of the simplest means to gather information about the potential function of a receptor is to determine its anatomical distribution. Analysis of the expression of the SSTR subtypes in various tissues has been ongoing since the identification of the receptors.[16-27] Results from these studies, however, reveal several problems with the concept that a given biological function is regulated by a single, specific SSTR subtype. First, most potential target tissues express not just one, but multiple SSTR subtypes.[22,26] Second, each of the five SSTR subtypes is widely distributed across multiple tissues.[16,17,20,21,23-25] Third, and presenting one of the major reasons why receptor expression results often appear variable, SSTR subtypes that are expressed by a particular tissue change over time in response to the physiological or pathological milieu.

The dynamic nature of SSTR receptor subtype expression is exemplified by the study of Khare et al,[27] which examined changes in SSTR subtype expression in rat aorta following injury. In this study, uninjured aorta was observed to express high levels of SSTR3, moderate levels of SSTR1 and 4, and very low levels of SSTR2 and 5. SSTR1 and SSTR2 expression increased significantly 3 and 7 days following injury, and returned to baseline by 14 days. In contrast, the expression of SSTR3 and 4 gradually increased over the entire study period following injury. No change was observed in SSTR5 expression. These results demonstrate how the pattern of SSTR expression can change over time in response to a specific challenge.

A clear example of how the prevailing hormonal milieu dictates the expression of SSTR subtypes is found in the elegant study of Visser-Wisselaar et al.[28] In this study, it was observed that when rat 7315b prolactinoma cells were maintained in estrogen-containing medium, they expressed both SSTR 2 and 3, and responded to Octreotide, a potent SSTR2-preferring SST analog, by decreasing prolactin secretion. When grown in the absence of estrogen, however, expression of SSTR2 and 3 was lost, and the cells became unresponsive to Octreotide. When tumors from the 7315b prolactinoma cells were grown in nude mice, elevated prolactin secretion from the tumors significantly suppressed the secretion of estradiol produced by the ovary, resulting in a loss of SSTR expression by the tumor and, consequently, the ability of the tumor to respond to Octreotide. Exogenous replacement of estrogen restored the expression of SSTR2 in the prolactinoma cells, and restored the ability of the tumor cells to bind Octreotide. These observations have clear implications for the use of SST analogs in the treatment of hyperprolactinemia, and strikingly demonstrate the impact of the prevailing hormonal milieu on the expression of SST receptor subtypes and the resulting ability of specific tissues to respond to SST.

To approach the problem of elucidating which SSTR subtypes are involved with a particular biological action of SST, we chose a functional approach in which we tested SST analogs with differing activity at the various SSTR subtypes in well-characterized assays for a specific biological function. At first, the analogs used were best described as SSTR subtype-preferring, in that the majority of activity was directed at one receptor subtype, but moderate to significant activity remained at one or more additional subtypes. Later analogs were refined to become receptor subtype selective, such that only one specific subtype was recognized. Whenever possible, the selected biological models were based on human tissues or cells in order to avoid species-related differences in SSTR function and/or structure.

The functional assays yielded clear associations of specific SSTRs with biological actions, such as SSTR5 directly suppressing insulin secretion from pancreatic β-cells,[15] and SSTR1 directly inhibiting new vessel outgrowth in human umbilical vein explants;[29] however, after examining multiple models, it became apparent that there is usually more than one SSTR subtype involved in a specific biological response. It also become apparent that the SSTR subtypes involved in a response often interact to produce a response that is distinct from that achieved by activation of an individual receptor subtype. Taken together, the observation that the ultimate biological response of a tissue to SST is comprised of the interplay of multiple SSTR subtypes, and that the SSTR subtypes that are available for interaction depend on the physiological/pathological status of the animal, suggest a mechanism by which very precise tissue responses can be generated according to the prevailing need. The remaining sections of this chapter will illustrate examples of this principle.

Somatostatin Receptor Subtypes in the Suppression of GH Secretion—First Level of Complexity

As indicated previously, suppression of GH secretion was the physiological action upon SST was first identified and isolated, and it remains the most clearly recognized function of SST. Despite this, the analogs of SST that are currently available for clinical use, Lanreotide and Octreotide, are not fully efficacious in normalizing GH in all patients with excess GH secretion. These analogs normalize GH in only approximately 50% of acromegalic patients.[30] It is recognized that 5-10% of acromegalics do not respond to SST at all, due to defects in either the SSTRs or in the signal transduction systems. The remaining 40-45% of acromegalic patients respond to, but are not fully normalized by, SST therapy.

As mentioned earlier, both Lanreotide and Octreotide were optimized on the basis of suppressing GH secretion in rat-based assays. With the elucidation of the SSTR subtypes, it was found that there is a very high correlation between the activity of SST analogs at the SSTR2 subtype and the ability to suppress GH from rat anterior pituitary cells.[31] It is, therefore, not surprising that Lanreotide and Octreotide are potent SSTR2-prefering compounds. However, we wondered if there might be a fundamental difference between the regulation of GH secretion in rats and humans. To attempt to answer this question, in collaboration with Ilan Shimon and Sholmo Melmed at Cedars-Sinai Medical Center (Los Angeles, CA, USA), we examined the ability of a panel of SSTR subtype-selective SST analogs to suppress GH secretion in cultured cells from both normal human fetal pituitary and from human GH-secreting adenomas. Just as in the rat, activation of the SSTR2 subtype suppresses GH secretion in human pituitary cells; however, unlike the rat, activation of the SSTR5 subtype in the human cells also suppresses GH secretion.[32] Even more intriguingly, it was observed that when the SSTR2 and 5 subtypes are coactivated, the suppression of GH is dramatically increased (Fig. 1).[33] Because natural SST is capable of activating all five known SST subtypes, it seems logical to assume that the enhanced suppression of GH induced by coactivating SSTR2 and 5 represents the normal mechanism by which SST suppresses GH in humans. This may provide one reason for the partial efficacy of the current SSTR2-preferring SST analogs in normalizing GH in the acromegalic patient population.

In order to test whether greater activation of the SSTR5 subtype, in combination with activation of SSTR2, might be more efficacious in suppressing GH in acromegalic subjects, in collaboration with Alexandru Saveanu and Philippe Jaquet at the Centre National de la Recherche Scientifique, Universite de la Mediterranee (Marseilles, France), we examined the effect of combined SSTR2 and SSTR5 activation on GH secretion from cultured pituitary adenoma cells collected from acromegalic patients who were classified as only partially responsive to SST analog therapy. As illustrated in Figure 2, the cultured, GH-secreting adenoma cells mimic the clinical response of the donors by responding only partially to addition of Octreotide to the culture medium.[34] Treatment with individual SST analogs that are potent and selective for either SSTR2 or SSTR5 yields a slightly greater increase in GH suppression. However, treatment with a combination of SSTR2 and 5 analogs produces a significant increase in GH suppression, comparable to that observed with cultured adenoma cells collected from patients classified as fully-responsive to SST therapy (Fig. 2).[34] These results strongly support the hypothesis that insufficient activation of the SSTR5 subtype by the current generation of clinical SST analogs may be one of the reasons why they are only partially effective in normalizing GH in a percentage of acromegalic patients.

To further extend these findings, we developed an analog of SST with potent, selective affinity for both the SSTR2 and 5 subtypes. The bi-selective SSTR2 + 5 analog, BIM-23244, is twice as potent as either Lanreotide or Octreotide at the SSTR2 subtype, and approximately 20-fold more potent at the SSTR5 subtype. To test the effect of this dual, receptor analog on GH secretion, again, in collaboration with Alexandru Saveanu and Philippe Jaquet, tumors from 10 acromegalic patients, 5 of whom were classified as fully responsive to Octreotide, and 5 of whom were classified as only partially responsive to Octreotide, were placed in culture and

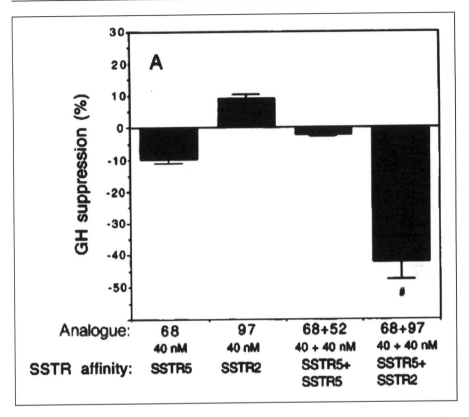

Figure 1. Example of interaction between SSTR2 and SSTR5 preferring SST analogs yielding enhanced suppression of GH secretion from cultured human GH-secreting adenoma cells. Concentrations of individual SST2 (BIM-23197) or SSTR5 (BIM-23268 and BIM-23052) were utilized that, alone, yielded minimal or no effect on GH secretion. Results are expressed as percent GH level versus that of untreated control wells. Each bar and bracket represents the mean + SEM of 8 replicate wells incubated with the indicated treatment for 3 hours in serum-free, defined medium. Reprinted with permission from Shimon et al.[33] Copyright 1997, The American Society for Clinical Investigation.

exposed to a range of doses of either Octreotide or BIM-23244. As anticipated from prior work, Octreotide produced a full, dose-related suppression of GH secretion in cultures of adenoma cells from patients classified as fully responsive to SST therapy, but only partially suppressed GH secretion in cultures of adenoma cells from patients that were partially responsive to SST therapy. In contrast, BIM-23244 induced a full, dose-related suppression of GH in all of the cultured adenomas, regardless of the donor's clinical response to Octreotide (Fig. 3).[35] These results again support the concept that full suppression of GH secretion in humans requires dual activation of the SSTR2 and 5 subtypes, and support the idea that SST analogs that can more fully activate both SSTR2 and 5 should be able to normalize GH secretion in a greater number of acromegalic patients than the SST analogs currently used clinically.

While these observations suggest functional interactions between SSTR subtypes, they do not suggest a mechanism by which these interactions occur. It may be as simple as activation of two systems with a common endpoint, or greater activation of intracellular or complimentary intracellular signal transduction systems. There are recent suggestions, however, that there may also be a physical interaction between the receptors. It has been observed that members of

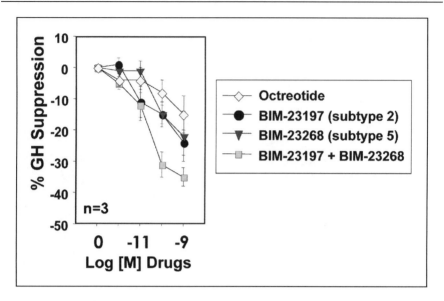

Figure 2. Combined treatment with SSTR2 (BIM-23197)- and SSTR5 (BIM-23268)-preferring analogs induces enhanced suppression of GH secretion from GH-secreting pituitary adenoma cells collected from an acromegalic patient classified as only partially responsive to the SST analogs currently available clinically. Use of the SST2 and SSTR5-preferring analogs individually produced a partial suppression of GH secretion, in keeping with the clinical profile of the tumor. Results are expressed as mean + SEM percent GH secretion vs. that of untreated control wells (medium alone). Reprinted with permission from Jaquet et al.[34] Copyright 2000, The Endocrine Society.

certain G-protein-coupled receptor (GPCR) families appear to form homo- and heterodimeric receptor combinations.[36,37] This phenomenon was observed with the κ and δ opiate receptors by Jordan and Devin,[38] who found that formation of heterodimers resulted in unique ligand interaction, as well as potentiated signal transduction. The formation of heterodimers between receptor subtypes has also been observed for the SSTRs. Rocheville et al[39] reported heterodimer formation between SSTR subtypes 1 and 5, with resulting enhanced signal transduction. If a similar phenomenon occurs between SSTR subtypes 2 and 5 on pituitary somatotrophs, it may provide an explanation for the enhanced suppression of GH observed when these receptors are coactivated. In addition, if receptor dimerization represents a naturally occurring physiological mechanism, then it could provide an additional means by which functional specificity for a widely distributed factor, such as SST, is achieved. In this scenario, the target tissue dictates the response to SST by expressing different ratios of SSTRs in response to the prevailing physiological and environmental conditions. It also provides a potential mechanism for further increasing the functional therapeutic specificity of SST analogs, since an analog that specifically interacts with only the dimerized combination of certain receptors should only impact those cells expressing that particular receptor combination.

Somatostatin Receptor Subtypes in the Suppression of Thyroid Medullary Carcinoma Proliferation—Further Complexity

Adding further complexity to the interplay between the SSTR subtypes is the observation that certain combinations of SSTR subtypes are antagonistic. This concept further expands the opportunities for a target cell to exert control over its response to SST and, by extension, provides new opportunities for target-specific SST analogs. Continuing our efforts to determine

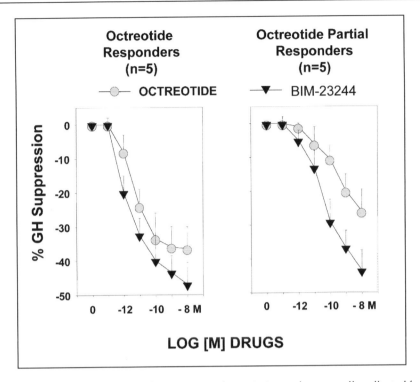

Figure 3. Enhanced suppression of GH secretion from pituitary adenoma cells collected from acromegalic patients classified as only partially responsive to Octreotide therapy. Samples of GH-secreting adenomas were collected at surgery from 10 acromegalic patients, 5 of whom were classified as fully responsive, and 5 of whom were classified as only partially responsive to Octreotide therapy, based on the results of a prior acute Octreotide challenge. The cultured adenoma cells were exposed to either Octreotide or BIM-23244, a bi-selective SST analog with potent activity at both SSTR2 and SSTR5. Octreotide and BIM-23244 induced a comparable, dose-related suppression of GH secretion from the adenoma cell cultures collected from patients classified as fully responsive to Octreotide. Addition of Octreotide to the adenoma cells cultures collected from patients classified as only partially responsive to Octreotide, produced only a partial suppression of GH; however, BIM-23244 remained able to induce a dose-related suppression of GH comparable to that induced in the adenoma cells from Octreotide fully responsive patients. Results are expressed as mean ± SEM percent GH suppression vs. medium alone (control). Reprinted with permission from Saveanu et al.[35] Copyright 2001, The Endocrine Society.

which SSTRs are associated with specific functions, we collaborated with Maria Chiara Zatelli and Ettore degli Uberti at the University of Ferrara (Ferrara, Italy) to determine the SSTR receptor subtypes that may be involved in regulating the calcitonin-secreting, parafolicular cell-derived tumors of the thyroid, thyroid medullary carcinoma (TMC). Using the human TMC cell line, TT, it was found that these cells express all five of the SSTR subtypes.[26] When tested with a panel of SSTR subtype-specific SST analogs, it was observed that activation of SSTR2 suppressed proliferation while activation of SSTR5 was without effect. In addition, selective activation of SST2 suppressed, while activation of SSTR5 stimulated, thymidine incorporation into the cellular DNA. The anti-proliferative action of the SSTR2 agonists was confirmed by demonstrating that the effect could be blocked using an SSTR2-selective antagonist. In the presence of a sufficient concentration of a SSTR2-seletive agonist to significantly suppress both thymidine

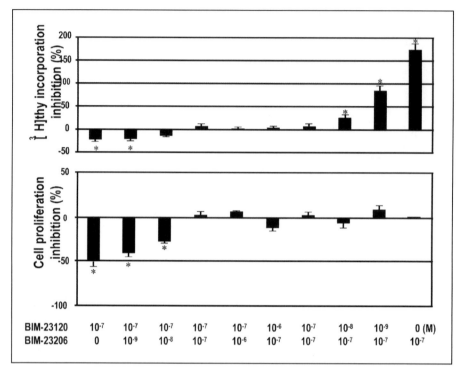

Figure 4. Progressive antagonism of SSTR2-induced suppression of [^{3}H]thymidine incorporation (upper panel) and proliferation (lower panel) by increasing activation of SSTR5 in the thyroid medullary carcinoma cell line, TT. Conversely, SSTR5-stimulated [3H]thymidine incorporation is progressively antagonized by increasing activation of SSTR2. Activation of SSTR2 was via addition of the SSTR2 selective SST agonist, BIM-23120, and activation of SSTR5 was via addition of the SSTR5-selecive SST agonist, BIM-23206. Upper Panel: Data from 6 individual experiments containing 4 replicates each, expressed as the mean ± SEM percent inhibition of [^{3}H]thymidine incorporation as compared with vehicle-treated controls. *, $P < 0.05$ and **, $P < 0.01$ vs. control. Lower Panel: Data from 6 individual experiments containing 8 replicates each, expressed as the mean ± SEM percent inhibition of cell proliferation as compared with vehicle-treated controls. *, $P < 0.05$ and **, $P < 0.01$ vs. control. The results were analyzed by ANOVA, followed by Student's paired or unpaired t tests to determine differences between individual means. Reprinted with permission from Zatelli et al.[26] Copyright 2001, The Endocrine Society.

incorporation and proliferation, addition of increasing amounts of a SSTR5-selective agonist resulted in a concentration-dependent ameliorization of the SSTR2 effect until, at equimolar concentrations, both the SSTR2-induced suppression of thymidine incorporation and proliferation were abolished (Fig. 4). Conversely, in the presence of a concentration of an SSTR5 agonist sufficient to significantly stimulate thiamine incorporation, the effect could be gradually abolished by increasing concentrations of an SSTR2-selective agonist (Fig. 4).[26]

These results are striking for several reasons. First, they illustrate that interactions between SSTR subtypes can be not only synergistic, but also antagonistic. This finding may explain, at least in part, the "bell-shaped" dose-response curves that are often observed with native SST and analogs that have activity at multiple SSTR subtypes. At low concentrations the most sensitive, dominant effect is observed. As the concentration is increased, however, additional receptors that are antagonistic to the primary response may be activated and gradually diminish the response. Second, the results in TMC demonstrate that a gradated response can be achieved depending on the ratio of SSTR subtype activity. This suggests that the target cells can, in a very precise way, dictate not only the type of response to SST, but can finely control the magnitude of the response. Third, the observation that the negative interaction involved the same SSTR subtypes that are synergistic in suppressing GH secretion from the pituitary further demonstrates the tissue-specific nature of SST physiology, and that control is local rather than systemic. Finally, the ability to target receptor interactions that are antagonistic for certain biological responses, in addition to those interactions that are synergistic for certain biological responses, provides a further means of enhancing the functional specificity of SST analogs by being able to target and suppress unwanted side effects.

A further observation from these studies in TMC is that while neither SSTR2- nor SSTR5-selective agonists are able to suppress calcitonin secretion, SSTR1-selective agonists suppress both calcitonin secretion and proliferation.[40] Interestingly, selective SSTR1 agonists appear to be more efficacious in these functions than natural SST, which produces a "bell-shaped" dose-response curve in this system (Fig. 5). As suggested previously, this phenomenon may be due to the activation of antagonistic receptor combinations by higher concentrations of native SST, which is able to activate all five of the known SSTR subtypes. The selective SSTR1 agonists display a sigmoidal dose-response curve that reaches and maintains a maximal plateau at higher concentrations (Fig. 5).

Since these original observations, evidence has emerged that like the synergistic effect of SSTR combinations, antagonistic interactions may also be due to dimerization of the SSTRs. In a report from Pfeiffer et al,[41] it was demonstrated that SSTR3 can interact with SSTR2 to form a heterodimeric complex that results in a loss of activity.

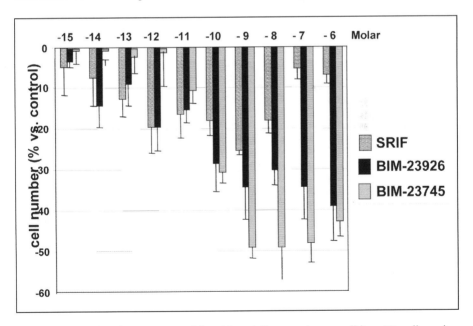

Figure 5. Dose-related suppression of thyroid medullary carcinoma cell line, TT, cell number following exposure to native SST and two SSTR1-selective SST analogs, BIM-23745 and BIM-23926. Both SST1-selective analogs reach and maintain a plateau of suppression at higher concentrations, whereas SST produces a "bell-shaped" curve in which the suppressive effect is lost at higher concentrations. Control wells were treated with vehicle. TT cell number was measured as absorbance at 490 nM of each well. Data from six individual experiments, with eight replicates each, are expressed as mean ± SEM percent cell proliferation inhibition versus vehicle-treated control cells. *P < 0.05 and **P < 0.01 vs. control.

Expanding beyond SSTR Subtypes—Interactions between SST and Dopamine Receptors

Another recent finding that further expands the complexity of the SST system is the demonstration that not only are there functional and, perhaps, physical interactions between the subtypes of a particular receptor family, such as the opiates or SST, there are also interactions between the members of different receptor families. Directly related to the present discussion of GH regulation is the interaction between the SSTR subtypes and members of the dopamine (DA) receptor family. Suggestions of potential interactions first emerged from the clinical literature in which it was reported that combined treatment of acromegalic patients with SST and DA agonists produced greater suppression of GH secretion then either agent alone.[42-45] One of the more detailed studies suggested that the effect was due to alterations of the pharmacokinetic properties of the compounds since it was observed that oral DA agonists were ~30% more bioavailable in the presence of SST agonist treatment.[42] Evidence for direct interaction at the cellular or receptor level, however, remained to be elucidated.

To more fully explore this interaction, in collaboration with Song-Guang Ren and Shlomo Melmed at Cedars-Sinai Medical Center (Los Angeles, CA, USA), we observed that both SST and DA agonists were able to suppress GH secretion from rat anterior pituitary cells. When tested in combination, greater potency was observed in suppressing GH secretion from the rat pituitary cells, but suppression of GH was only to the same maximal extent as observed with the individual agonists, i.e., no greater efficacy in suppressing GH (unpublished observations). Similarly, in collaboration with Philippe Jaquet's group at the Centre National de la Recherche Scientifique, Universite de la Mediterranee (Marseille, France), we examined the effect of combined SST and DA agonist treatment on prolactin secretion from human prolactinomas. In agreement with previous studies that demonstrated that SSTR5 suppresses prolactin secretion from human prolactinoma cells,[33] it was observed that both SSTR5 and DA agonists suppressed prolactin secretion in a dose-dependent manner, and as with GH secretion in the rat, combined treatment results in greater potency, but not in greater efficacy.[46] The results from these two studies suggest direct functional interaction between SST and DA receptors at the cellular level. In support of a possible direct physical interaction, heterodimer formation between the DA D2 receptor and both SSTR5[47] and SSTR2[48] has been described.

In an effort to capture the enhanced potency that we observed with the SST/DA combination within a single compound, we have produced chimeric molecules that possess structural elements of both SST and DA, and that retain the ability to bind to and to activate both the SSTR2 and the DA D2 receptors.[49] Again, working with Philippe Jaquet's group in Marseille, these SST/DA chimeras were tested for their ability to suppress GH secretion from cultured human GH-secreting adenoma cells collected from patients who were classified as only partially responsive to the current clinically used SST analogs. As expected, individual SSTR2 and DA D2 activating ligands were able to suppress GH in a dose-related manner (Fig. 6A). In contrast to the clinical literature, however, adding the two individual SSTR2 and DA D2 agonists in combination produced no greater suppression of GH than the addition of the SST agonist alone (Fig. 6A). Despite the lack of functional interaction with the individual ligands, the chimeric molecule, with dual SST2 and DAD2 activity, displayed an approximately 100-fold increase in potency, with efficacy at least equal to that of the best SST analogs (Fig. 6B).[50]

Combining the information from our studies demonstrating functional interaction between SSTR2 and SSTR5, and between SSTR2 and DA D2, we began comparing the GH-suppressing efficacy of chimeras possessing varying ratios of activity at all three receptors, SSTR2, SSTR5 and DA D2. We observed considerably greater efficacy in suppressing GH secretion with certain ratios of these three activities as compared with individual SST or DA analogs, while the tremendously increased potency observed in the original chimera studies was retained. We observed that the greatest efficacy was achieved with chimeric compounds displaying either potent SSTR2, potent SSTR5 and moderate DA D2 activity, or

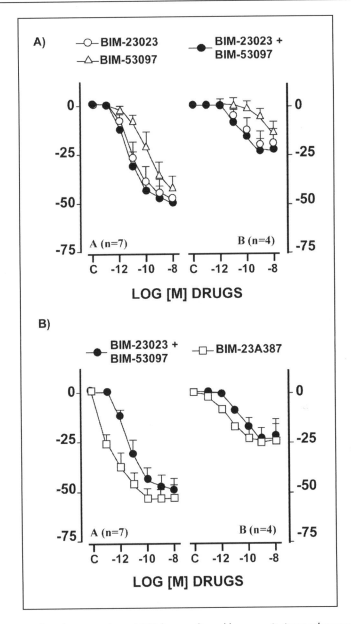

Figure 6. Dose-related suppression of GH from cultured human pituitary adenoma cells collected from patients classified as fully- (left panel, A and B), or only partially- (right panel, A and B) responsive to Octreotide therapy. Panel A) Comparison of the potency and efficacy of the SSTR2-preferential SST analog, BIM-23023, and the DA D2 agonist, BIM-53097, both alone and in combination. Panel B) Comparison of the potency and efficacy of the combination of BIM-23023 and BIM-53097, with the SST/DA chimeric molecule, BIM-23A387. Results are expressed as the mean + SEM percent GH suppression versus vehicle-treated controls. Each point represents four replicate wells. Reprinted with permission from Saveanu et al.[50] Copyright 2002, The Endocrine Society.

potent SSTR2, moderate SSTR5 and potent DA D2 activity.[51] Interestingly, compounds with high activity at all three receptors produced GH-suppression that was merely comparable to that of the current clinical SST analogs.[51] The reason for this phenomenon is presently unknown. If, however, the enhanced activity of these chimeras does indeed stem from physical interaction (dimerization) between the receptors, there may be steric interference created with a compound that can potently bind more than two interacting receptors.

One of the most potent and efficacious of the chimeric molecules is BIM-23A760, which has binding affinity (Ki) of 0.03, 42, and 25 for the SSTR2, SSTR5 and DA D2 receptors, respectively. Recently, we tested the ability of BIM-23A760 to suppress GH secretion in normal cynomolgus monkeys (Macaca fascicularis).[52] The study was designed as an escalating dose comparison with single, subcutaneous injections ranging from 0.001 to 1000 nmole/kg, and with a 10 day wash out period between each dose. BIM-23A760 produced potent, significant dose-related suppression of GH that was evident within 15 minutes and through 48 hours after injection. The dose of BIM-23A760 required to induce a 50% decrease in the highest preinjection GH level was 0.56nmole (~1 µg)/kg at 2 hours and 10nmole (~17 µg)/kg at 24 hrs post-injection. Dose-related suppression of prolactin was also observed. The observed efficacy of BIM-23A760 in suppressing GH was somewhat unexpected, since it is known that in normal animals, SST suppresses, while DA stimulates, GH secretion. Because of these potentially opposing actions, it is anticipated that even though impressive efficacy was observed in the normal primates, BIM-23A760 may be even more potent and efficacious in acromegalic patients, in whom GH secretion is suppressed by both SST and DA.

The second intriguing observation from the cynomolgus study is that BIM-23A760 had no effect on circulating insulin or glucose levels.[53] In contrast, under the same conditions, a SST analog with affinity less than 10nM for all of the SSTR subtypes except SSTR4, induced dose-related suppression of both insulin and glucose. Interestingly, the dopamine agonist, Pergolide, under the same experimental conditions, produced a dose-related increase in circulating glucose without impacting insulin secretion. These results suggest that chimeric SST/DA molecules such as BIM-23A760 may provide effective therapy for a greater percentage of acromegalics than the current clinical SST and DA analogs, without significantly affecting pancreatic function and glycemic control.

SST-DA Interactions—How Does It Work and What Does It Mean?

The findings described in the previous section collectively illustrate the enhanced efficacy that can be achieved through concurrent activation of SST and DA receptors with a single molecule possessing both SST and DA activities. While the mechanism(s) by which this effect is achieved remains to be elucidated, several possibilities have been proposed. The first, as already mentioned, is the formation of heterodimers between the SST and DA receptors resulting in altered binding and enhanced signal transduction. Evidence consistent with receptor dimerization in the presence of a chimeric SST/DA molecule has been generated by Anat Ben-Shlomo and Shlomo Melmed at Cedars-Sinai Medical Center (Los Angeles, CA, USA). Fluorescently tagged antibodies generated against SSTR2 and DA D2 were used to demonstrate that these receptors are present on the surface of the same cells of the mouse corticotroph cell line, AtT20. Using a technique known as Fluorescence Resonance Energy Transfer or FRET, the two antibodies were tagged with either acceptor or donor fluorophores, such that photonic energy is exchanged between the fluorophores depending on the physical proximity of the two antibodies to each other. Under basal conditions, the FRET value was very low, suggesting that the SST and DA receptors present on the AtT20 cells were not in close physical proximity. When a chimeric SST/DA molecule was introduced to the system, however, the FRET value increased dramatically, indicating that the two labeled antibodies, and by extension, the SSTR2 and DA D2 receptors, were now in close physical proximity (unpublished observations). When a DA antagonist, sulpiride, was added to the mixture to impair binding to the DA D2 receptor, the FRET value returned to the pretreatment, basal level. While not definitive, these results are consistent with the theory that the receptors form heterodimers in the presence of the chimeric SST/DA molecule. How the chimera might induce dimerization remains a mystery. It seems unlikely that the chimeras, which are not much larger than the octapeptide analogs, Lanreotide and Octreotide, are able to bind across two separate receptor binding domains. It may be possible that binding of a ligand to either one or both of the receptors induces conformational changes that facilitate dimerization, and that the combined receptors form a unique binding site that is optimal for the chimera, but this possibility and other speculations must await additional data.

Another possibility is that the chimeric molecules may achieve greater potency by inducing more protracted receptor activation, either by slowing or preventing receptor internalization, thus keeping the receptor available for a longer period, or by providing greater opportunity for the ligands to interact with the receptors. If it is true that ligands can bind to a receptor, dissociate, and bind again to another receptor, then the chimeras offer the possibility of greater receptor activation simply by increasing the probability of interacting with available receptors of multiple types. Increased receptor interaction may also keep the chimera in close association with the cell membrane for a longer period of time, thus further facilitating additional receptor interactions. Evidence that the chimeras may induce a more protracted cellular response was produced by the group of Philippe Jaquet at the National de la Recherche Scientifique, Universite de la Mediterranee (Marseilles, France). A SSTR2 agonist, a DA D2 agonist, or one of the chimeric molecules was individually added to the medium of cultured human GH-secreting pituitary adenoma cells. After 30 minutes, the compounds were washed off, replaced with fresh media alone, and the subsequent changes in GH secretion were monitored. Maximal suppression of GH was observed 6 hours after exposure to either the SSTR2 or the DA D2 agonist. This effect was maintained for at least 12 hours, but was lost by 24 hours after compound exposure. In contrast, treatment with the chimeric molecule induced a much more rapid suppression of GH, reaching maximal suppression within 3 hours, and achieving a level of GH suppression significantly greater than either of the SST or DA ligands alone. The chimera-induced suppression of GH lasted at least 24 hours and was lost by 36 hours after compound exposure (unpublished observations). These results demonstrate that interaction between the receptor and the SST/DA chimeras results in both a greatly accelerated, as well as protracted, biological response. Unfortunately, once again the mechanism responsible for the unique time course of SST-DA chimera action requires further elucidation.

Beyond acromegaly, the unique action of the SST/DA chimeras may also provide effective therapy for other conditions. The chimeric SST/DA molecules have already been demonstrated to be highly efficacious and far more potent in suppressing prolactin secretion from mixed GH-prolactin secreting adenoma cells than individual DA agonists (Fig. 7).[50,51] Consequently, these chimeras may prove useful in treating prolactinomas that are resistant to the DA analogs currently used clinically. There are also limited clinical reports suggesting that combined therapy with SST and DA agonists may offer greater efficacy in suppressing adrenocorticotropic hormone (ACTH) secretion from pituitary corticotroph tumors (Cushing's disease),[54] and in suppressing growth of nonfunctioning pituitary adenomas (NFPAs: usually gonadotroph in origin).[55] Initial reports have demonstrated that the SST/DA chimeras are effective in suppressing proliferation of cultured human NFPA cells.[56,57] In addition, dopamine agonists are highly efficacious in inducing shrinkage of prolactinomas,[58] while the efficacy of SST agonists in shrinking GH-secreting adenomas is more variable.[59] It is possible that the combined, enhanced dopaminergic and somatostatinergic activity of the chimeric molecules may provide greater efficacy and consistency in inducing shrinkage of multiple types of adenomas. Studies are currently ongoing to assess the potential of the chimeras as therapeutic agents for pituitary tumor types beyond GH-secreting adenomas.

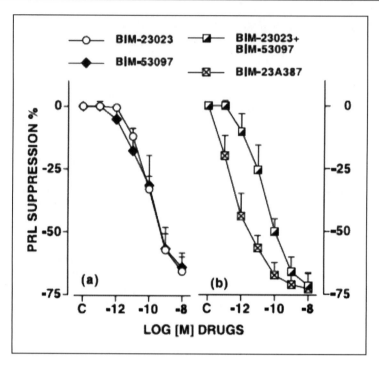

Figure 7. Comparison of the dose-related suppression of prolactin secretion from cultured, mixed GH/prolactin-secreting human pituitary adenoma cells by individual SSTR2 (BIM-23023) and DA (BIM-53097) agonists, both alone (left panel) and in combination (right panel), and the SSTR/DA chimeric molecule, BIM-23A387 (right panel). Results are expressed as mean + SEM percent prolactin secretion versus control (medium alone) from 6 different adenomas. Each point and bracket represents the results from four replicate wells. Reprinted with permission from Saveanu et al.[50] Copyright 2002, The Endocrine Society.

Opportunity from Complexity

The foregoing observations regarding the interaction between SSTR subtypes, and between members of different receptor families, demonstrate the complexity of the interplay between ligand and receptor, and the potential for regulation of the cellular response. If G-protein-coupled receptors from different families can combine, either physically, functionally or both, to enhance or suppress cellular responses, or to create a novel response, then the types and ratio of different receptors present may dictate the cellular response to incoming hormonal signals. Because the expression and level of expression of each receptor type is determined by the prevailing physiological, pathological and environmental conditions, and because the factors regulating each of the different receptor families differs, the potential for different cellular responses, and magnitudes of individual cellular responses is enormous. This degree of complexity and range of responses provides a tremendous degree of flexibility for the cell to respond to moment to moment changes in the conditions encountered by the parent organism.

The regulation of cellular response through the interaction of different receptors provides a new concept for modeling disease states. Through any number of imaginable scenarios that could influence receptor expression, a cell or tissue could lose the ability to express the proper ratio of receptors needed to maintain homeostasis. Conversely, the interplay of different receptor species that are associated with a specific cell-type in a specific tissue creates the possibility of creating drugs that are targeted to interact with only a specific receptor combination that has been generated in response to a specific condition. This would seem to be the ultimate key to achieving therapeutic specificity, as well as maximal therapeutic efficacy. Our challenge for the future is to unravel the complex receptor coding that dictates cellular response.

References

1. Brazeau P, Vale W, Burgus R et al. Hypothalamic polypeptide that inhibits the secretion of immunoreactive pituitary growth hormone. Science 1973; 179:77-9.
2. Hofland LJ, Visser-Wisselaar HA, Lamberts SWJ. Somatostatin analogs: Clinical application in relation to human somatostatin receptor subtypes. Biochemical Pharmacology 1995; 50:287-97.
3. Reichlin S. Somatostatin (Part 1). N Engl J Med 1983; 309:1495-501.
4. Reichlin S. Somatostatin (Part 2). N Engl J Med 1983; 309:1556-563.
5. Sheppard M, Shapiro B, Pimstone B et al. Metabolic clearance and plasma half-disappearance time of exogenous somatostatin in man. J Clin Endocrinol Metab 1979; 48:50-53.
6. Grant N, Clark D, Garsky V et al. Dissociation of somatostatin effects. Peptides inhibiting the release of growth hormone but not glucagon or insulin in rats. Life Sci 1976; 19:629-31.
7. Meyers C, Arimura A, Gordin A et al. Somatostatin analogs which inhibit glucagon and growth hormone more than insulin release. Biochem Biophys Res Commun 1977; 74:630-6.
8. Brown M, Rivier J, Vale W. Somatostatin: Analogs with selected biological activities. Science 1977; 196:1467-9.
9. Coy DH, Meyers C, Arimura A et al. Observations on the growth hormone, insulin, and glucagon release-inhibiting activities of somatostatin analogues. Metabolism 1978; 27(9 Suppl 1):1407-10.
10. Bauer W, Briner U, Doepfner W et al. A very potent and selective octapeptide analogue of somatostatin with prolonged action. Life Sci 1982; 31:1133-40.
11. Heiman ML, Murphy WA, Coy DH. Differential binding of somatostatin agonists to somatostatin receptors in brain and adenohypophysis. Neuroendocrinology 1987; 45:429-36.
12. Reisine T, Bell GI. Molecular biology of somatostatin receptors. Endocr Rev 1995; 16:427-42.
13. Coy DH, Murphy WA, Raynor K et al. The new pharmacology of somatostatin and its multiple receptors. J Pediatr Endocrinol 1993; 6:205-9.
14. Rossowski WJ, Coy DH. Specific inhibition of rat pancreatic insulin or glucagons release by receptor-selective somatostatin analogs. Biochem Biophys Res Commun 1994; 205:341-346.
15. Zambre Y, Ling Z, Chen MC et al. Inhibition of human pancreatic islet insulin release by receptor-selective somatostatin analogs directed to somatostatin receptor subtype 5. Biochem Pharmacol 1999; 57:1159-64.
16. Bruno JF, Xu Y, Song J et al. Tissue distribution of somatostatin receptor subtype messenger ribonucleic acid in the rat. Endocrinology 1993; 133:2561-7.
17. Raulf F, Perez J, Hoyer D et al. Differential expression of five somatostatin receptor subtypes, SSTR1-5, in the CNS and peripheral tissue. Digestion 1994; 55(Suppl 3):46-53.
18. Reubi JC, Schaer JC, Waser B et al. Expression and localization of somatostatin receptor SSTR1, SSTR2, and SSTR3 messenger RNAs in primary human tumors using in situ hybridization. Cancer Res 1994; 54:3455-9.
19. Schonbrunn A, Gu YZ, Dournard P et al. Somatostatin receptor subtypes: Specific expression and signaling properties. Metabolism 1996; 45(8 Suppl 1):8-11.
20. Patel YC, Greenwood M, Panetta R et al. Molecular biology of somatostatin receptor subtypes. Metabolism 1996; 45(Suppl 1):31-8.
21. Reubi JC, Schaer JC, Markwalder R et al. Distribution of somatostatin receptors in normal and neoplastic human tissues: Recent advances and potential relevance. Yale J Biol Med 1997; 70:471-9.
22. Briard N, Dutour A, Epelbaum J et al. Species differences between male rat and ram pituitary somatostatin receptors involved in the inhibition of growth hormone secretion. Eur J Endocrinol 1997; 137:545-55.
23. Schonbrunn A. Somatostatin receptors present knowledge and future directions. Ann Oncol 1999; 10(Suppl 2):S17-S21.
24. Fehlmann D, Langenegger D, Schuepbach E et al. Distribution and characterisation of somatostatin receptor mRNA and binding sites in the brain and periphery. J Physiol Paris 2000; 94:265-81.
25. Kimura N, Schindler M, Kasai N et al. Immunohistochemical localization of somatostatin receptor type 2A in rat and human tissues. Endocrine J 2001; 48:95-102.
26. Zatelli MC, Tagliati F, Taylor JE et al. Somatostatin receptor subtypes 2 and 5 differentially affect proliferation in vitro of the human medullary thyroid carcinoma cell line TT. J Clin Endocrinol Metab 2001; 86:2161-9.
27. Khare S, Kumar U, Sasi R et al. Differential regulation of somatostatin receptor types 1-5 in rat aorta after angioplasty. FASEB J 1999; 13:387-94.

28. Visser-Wisselaar HA, Van Uffelen CJ, Van Koetsveld PM et al. 17-beta-estradiol-dependent regulation of somatostatin receptor subtype expression in the 7315b prolactin secreting rat pituitary tumor in vitro and in vivo. Endocrinology 1997; 138:1180-9.

29. Bocci G, Culler MD, Taylor JE et al. In vitro administration of novel and selective somatostatin subtype-1 receptor (SSTR-1) agonists inhibits three-dimensional proliferation of human endothelial cells. San Francisco, CA, USA: 84th Annual Meeting of the Endocrine Society, (Abstract #OR54-4).

30. Shimon I, Melmed S. Management of pituitary tumors. Ann Intern Med 1998; 129:472-83.

31. Raynor K, Murphy WA, Coy DH et al. Cloned somatostatin receptors: Identification of subtype-selective peptides and demonstration of high affinity binding of linear peptides. Mol Pharmacol 1993; 43:838-44.

32. Shimon I, Taylor JE, Dong JZ et al. Somatostatin receptor subtype specificity in human fetal pituitary cultures: Differential role of SSTR2 and SSTR5 for growth hormone, thyroid-stimulating hormone, and prolactin regulation. J Clin Invest 1997; 99:789-98.

33. Shimon I, Yan X, Taylor JE et al. Somatostatin receptor (SSTR) subtype-selective analogues differentially suppress in vitro growth hormone and prolactin in human pituitary adenomas: Novel potential therapy for functional pituitary tumors. J Clin Invest 1997; 100:2386-96.

34. Jaquet P, Saveanu A, Gunz G et al. Human somatostatin receptor subtypes in acromegaly: Distict patterns of messenger ribonucleic acid expression and hormone suppression identify different tumoral phenotypes. J Clin Endocrinol Metab 2000; 85:781-92.

35. Saveanu A, Gunz G, Dufour H et al. BIM-23244, A somatostatin receptor subtype 2- and 5-selective analog with enhanced efficacy in suppressing growth hormone (GH) from octreotide-resistant human GH-secreting adenomas. J Clin Endocrinol Meta 2001; 86:140-5.

36. Gouldson PR, Snell CR, Bywater RP et al. Domain swapping in G-protein coupled receptor dimers. Protein Eng 1998; 11:1181-93.

37. Hebert TE, Bouvier M. Structural and functional aspects of G protein-coupled receptor oligomerization. Biochem Cell Biol 1998; 76:1-11.

38. Jordan BA, Devin LA. G-protein-coupled receptor heterodimerization modulates receptor function. Nature 1999; 399:697-700.

39. Rocheville M, Lange DC, Kumar U et al. Subtypes of the somatostatin receptor assemble as functional homo- and heterodimers. J Biol Chem 2000; 275:7862-69.

40. Zatelli MC, Tagliati F, Piccin D et al. Somatostatin receptor subtype 1-selective activation reduces cell growth and calcitonin secretion in a human medullary thyroid carcinoma cell line. Biochem Biophys Res Commun 2002; 297:828-34.

41. Pfeiffer M, Koch T, Schroder H et al. Homo- and heterodimerization of somatostatin receptor subtypes - Inactivation of sst$_3$ receptor function by heterodimerization with sst$_{2A}$. J Bio Chem 2001; 276:14027-36.

42. Flogstad AK, Halse J, Grass P et al. A comparison of octreotide, bromocriptine, or a combination of both drugs in acromegaly. J Clin Endocrinol Metab 1994; 79:461-5.

43. Li JK, Chow CC, Yeung VT et al. Treatment of Chinese acromegaly with a combination of bromocriptine and octreotide. Aust N Z J Med 2000; 30:457-61.

44. Marzullo P, Ferone D, Di Somma C et al. Efficacy of combined treatment with lanreotide and cabergoline in selected therapy-resistant acromegalic patients. Pituitary 1999; 1:115-20.

45. Minniti G, Jaffrain-Rea ML, Baldelli R et al. Acute effects of octreotide, cabergoline and a combination of both drugs on GH secretion in acromegalic patients. Clin Ter 1997; 148:601-7.

46. Jaquet P, Ouafik L, Saveanu A et al. Quantitative and functional expression of somatostatin receptor subtypes in human prolactinomas. J Clin Endocrinol Metab 1999; 84:3268-76.

47. Rocheville M, Lange DC, Kumar U et al. Receptors for dopamine and somatostatin: Formation of hetero-oligomers with enhanced functional activity. Science 2000; 288:154-7.

48. Baragli A, Kumar U, Patel RC. Hetero-oligomerization of dopamine (D2R) and somatostatin receptors (SSTR2) in CHO-K1 cells and cortical cultured neurons. Philadelphia, PA, USA: Proceedings of the 85th Annual Meeting of the Endocrine Society, 2003; 469, (Abstract P2-669).

49. Culler MD, Taylor JE, Kim S et al. Chimeric analogs of somatostatin in pituitary adenoma. Paper presented at: "Somatostatin analogs: A state of the art review," a satellite symposium of the 5th European Congress of Endocrinology. Turin, Italy: 2001.

50. Saveanu A, Lavaque E, Gunz G et al. Demonstration of enhanced potency of a chimeric somatostatin-dopamine molecule, BIM-23A387, in suppressing growth hormone and prolactin secretion from human pituitary somatotroph adenoma cells. J Clin Endocrinol Metab 2002; 87:5545-52.

51. Jaquet P, Gunz G, Saveanu A et al. Efficacy of chimeric molecules directed towards multiple so-matostatin and dopamine receptors on inhibition of GH and prolactin secretion from GH-secreting pituitary adenomas classified as partially responsive to somatostatin analog therapy. Eur J Endocrinol 2005; 153:135-41.
52. Culler MD, Dong JZ, Taylor JE et al. The somatostatin-dopamine chimeric molecule, BIM-23A760, is highly efficacious in suppressing GH in normal, cynomolgus monkeys (Macaca fascicularis). Philadelphia, PA, USA: Proceedings of the 88th Annual Meeting of the Endocrine Society, 2006, (Abstract O9-6).
53. Culler MD, Dong JZ, Taylor JE et al. The somatostatin-dopamine chimeric molecule, BIM-23A760, does not induce the insulin/glycemic effects observed with individual somatostatin or dopamine agonists in cynomolgus monkeys (Macaca fascicularis). Athens, Greece: 12th Meeting of the European Neuroendocrine Association, 2006.
54. Pivonello R, Ferone D, Lamberts SWJ et al. Cabergoline plus lanreotide for ectopic Cushing's syndrome. N Engl J Med 2005; 352:2457-2458.
55. Anderson M, Bjerre P, Schroder HD et al. In vivo secretory potential and the effect of combina-tion therapy with octreotide and cabergoline in patients with clinically nonfunctioning pituitary adenomas. Clin Endocrinol 2001; 54:23-30.
56. Gruszka A, Kunert-Radek J, Radek A et al. The effect of selective sst1, sst2, sst5 somatostatin receptors agonists, a somatostatin/dopamine (SST/DA) chimera and bromocriptine on the "clini-cally nonfunctioning" pituitary adenomas in vitro. Life Sci 2006; 78:689-93.
57. Florio T, Barbieri F, Spaziante R et al. Dopamine/somatostatin chimeric molecules inhibit prolif-eration of human clinically nonfunctioning pituitary adenoma cells in vitro: A multi-center study. Athens, Greece: 12th Meeting of the European Neuroendocrine Association, 2006.
58. Molitch M. Medical management of prolactin-secreting pituitary adenomas. Pituitary 2002; 5:55-65.
59. Freda PU. Somatostatin analogs in acromegaly. J Clin Endocrinol Metab 2002; 87:3013-3018.

CHAPTER 4

Somatostatin Receptors in Human Tumors:
In Vitro Studies

Marek Pawlikowski*

Abstract

Neoplastic cells express and often even over-express the somatostatin receptors. It is important because the presence of sst receptors predicts—to some extent—the possibility of treatment with SST analogs. The presence of sst receptors may be examined in vivo by means of the receptor scintigraphy using the radiolabeled SST analogs. The receptors can be also detected in vitro (ex vivo) on surgical or biopsy specimens. Among the in vitro (ex vivo) methods, the immunohistochemical investigation with specific anti-receptor antibodies seems to be particularly useful for routine diagnostics. The chapter discusses the data on the incidence of sst1-5 receptor subtypes in the different human tumors including pituitary adenomas, brain tumors, thyroid and adrenal tumors, neuroendocrine tumors (NET), and nonendocrine cancers. These data lead to the conclusion that the expression of sst receptors is not restricted to the neuroendocrine and endocrine tumors but can be detected also in nonendocrine malignancies. The above suggests that the latter are also candidates for therapeutic trials with SST analogs. On the other hand, this possibility is limited by the fact that the detectable sst receptors are not always functional.

Introduction

Somatostatin (SST) acts via five different subtypes of receptors (sst1-5, for details see Chapter 2). These receptors are expressed, and often over-expressed in the tumoral tissues. The information on the state of expression of sst receptors in particular tumors is very important because it predicts the effectiveness of the therapy with SST analogs. The detection of sst receptors is possible under the in vivo and ex vivo-in vitro conditions. In vivo, the detection of sst receptors is allowed by receptor scintigraphy. This technique, presented in details in Chapter 7, is based on the linkage of SST analogs with a radionuclide. The ex vivo—in vitro techniques include autoradiography, molecular biology methods like reverse transcriptase—polymerase chain reaction (RT-PCR) or hybridization in situ (HIS) and immunohistochemistry. Each method discussed here has its advantages and disadvantages. In vivo receptor scintigraphy allows the precise localization of sst receptors-over-expressing tumor and its metastases within the patient's whole body. The investigation can be performed prior to surgery and in patients not submitted to surgical treatment. On the other hand, the investigation is practically limited to the receptor subtype sst2 because the available radiolabeled SST analogs are more or less selective ligands of sst2 receptor subtype. There is also no possibility to evaluate the precise localization of sst receptors within the tumor at tissue and cellular level. Autoradiography using the radiolabeled selective ligands for particular sst receptor subtypes is not

*Marek Pawlikowski—Department of Neuroendocrinology, Chair of Endocrinology, Medical University of Lodz, Sterling str 3, 91-425 Lodz, Poland. Email: pawlikowski.m@wp.pl

Somatostatin Analogs in Diagnostics and Therapy, edited by Marek Pawlikowski.
©2007 Landes Bioscience.

convenient for routine diagnostic applications because it is rather complicated and needs the freshly frozen material. Molecular biology techniques allow the precise characterization of the receptor subtypes. However, they present several limitations. They can be performed only ex vivo, using the material excised during surgery or obtained by biopsy. The evaluation of cellular localization of receptors is not possible. Because of their high sensitivity PCR can give the false positive results, due to the contamination of the material with some nontumoral cells expressing sst receptors like immune cells infiltrating tumor or vascular endothelial cells. Lastly, the expression at the level of mRNA is not always accompanied by the expression of the respective receptor protein, although in most cases the results are concordant. The advantages of immunohistochemical detection of sst receptors include the use of the same material which is used in routine pathological examination (paraffin sections), the possibility to detect all subtypes of sst1-5 and to localize precisely the receptor proteins within the examined tissue.

Incidence of Somatostatin Receptors in Human Tumors

Pituitary Adenomas

The normal human anterior pituitary gland is found to express all subtypes of sst receptors, except sst2B and sst4. However, the latter subtype is detectable in fetal pituitary tissue.[1] In pituitary adenomas the all subtypes are detectable at least in some cases, depending on the hormone phenotype of the tumors. GH-secreting adenomas expressed mostly sst2 (the distinction between sst2A and sst2B was not done in majority of studies) and sst5[1-4] (Fig. 1A). The data on sst1 and sst3 are variable from study to study, while sst4 subtype was not detected except for one study.[1] The majority of prolactinomas expresses sst1, sst2, and sst5, in approximatively one third of cases sst3 subtype was also detectable, while the expression of sst4 is very rare. In TSH-secreting adenoma all subtypes of sst receptors were found except sst2B.[1] The ACTH-secreting adenomas in patients with Cushing's disease sst2 and sst5 are present in majority of cases; sst1 and sst3 are less frequent (a half and one third of cases, respectively) and sst4 expression is exceptional. Recently, the importance of sst5 subtype in ACTH-secreting tumors is underlined. Subtypes sst2 and sst5 have different sensitivity to glucocorticoids, which down-regulate sst2 but not sst5 receptors. The above findings could explain the stronger inhibitory effect of SOM230, a SST analog having a higher affinity to sst5 subtype, in comparison to octreotide.[5] The nonfunctioning pituitary adenomas (NFPA) can express all subtypes of sst receptors. As in the other types of adenomas, the expression of sst4 is scarce. The remaining subtypes were found to be more abundant, but the data reported by different authors are not fully concordant.[1-4] Our own observations indicate that the dominating subtypes in NFPA are sst5 and sst2B, which are present in all the adenomas examined.[6] (Fig. 1B,C). Although NFPA represent the heterogenous group in respect to the expression of pituitary hormones or their subunits and only 1/5 of them are truly hormonally inactive, the only difference between the true inactive (null cell adenomas) and gonadotropin-expressing tumors concerned the appearance of sst4, which was more frequent in the former.[6] To summarize, pituitary adenomas frequently express sst receptors. The above constitutes a rationale to the application of SST analogs in order to suppress the pituitary hormone excess in case of functioning tumors and the growth inhibition of both functioning and nonfunctioning tumors. However, the treatment with SST analogs (octreotide and lanreotide) is a routine procedure only in the therapy of GH-secreting pituitary adenomas in patients suffering from acromegaly and of TSH-secreting pituitary adenomas (see Chapter 5 in this book). The trials of SST analogs in the treatment of the other types of pituitary adenomas are in progress, but the results are either controversial or limited in number.[7,8] Because the pituitary adenomas express often the subtypes of sst receptors different from sst2 (sst1, 3 and 5) we can lay some hope in the introduction to the routine therapy SST analogs acting on the above-mentioned receptors, e.g. SOM 230.

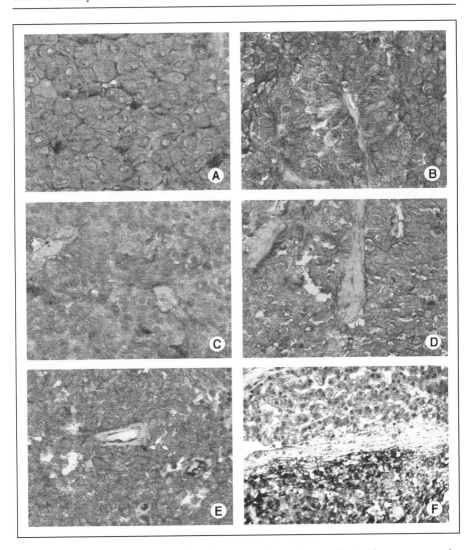

Figure 1. A) GH-secreting pituitary adenoma in the patient suffering from acromegaly. Immunostaining with anti-sst2A antibody (SS-800). The membranous localization of the immunoreaction. Reproduced from Pawlikowski M, Endokrynologia Polska 2005; 56:4-5, with permission of the editor. B) FSH/LH-secreting pituitary adenoma immunostained with anti-sst5 antibody (SS-890). C) Alpha-subunit -secreting pituitary adenoma immunostained with the anti-sst2B antibody (SS-860). B and C reproduced from Pawlikowski M et al. Endocrine Pathology 2003; 14:231-238, with permission of the editor. D) Neuroendocrine pancreatic cancer immunostained with anti-sst2A antibody (SS-800). Reproduced from Kunert-Radek J et al. Endokrynologia Polska Pol J Endocrinol 2004; 55:190-195, with permission of the editor. E) Neuroendocrine tumor (carcinoid) immunostained with anti-sst 5 antibody (SS-890). Positive immunostaining in the tumoral cells and vascular endothelia. F) Pheochromocytoma, metastasis to the liver. Strong immunostaining with the anti-sst3 antibody (SS-850).

Brain Tumors

High incidence of sst receptors in meningiomas was first reported using the autoradiography with radiolabeled SST and octreotide.[9] The more recent data, obtained with RT-PCR, revealed that 88% of investigated meningiomas express at least one of the five sst receptor subtypes. Subtypes sst1 and sst2 were the most frequently detected (69% and 79%, respectively); sst3, sst4 and sst5 were also present at least in one third of cases.[10] Somatostatin receptors are also detected in malignant brain tumors. Fruhwald et al[11] reported on immunostaining for sst2 receptors in children brain tumors and found the highest incidence in supratentorial primitive neuroectodermal tumors (7/7) The strong positive immunostaining was also found in 2/3 anaplastic astrocytomas and 2/3 anaplastic ependymomas. On the other hand, the reaction was low in 5 but 1 glioblastomas and 4 but 1 II grade ependymomas. High expression of sst2 receptor subtype is also described in medulloblastomas.[12] The data obtained in ex vivo studies do not predict the possibility of demonstration of the tumors in vivo by means of receptor scintigraphy, since the effectiveness of the latter depends also on the functional state of the blood-brain barrier.

Neuroendocrine Tumors

Under the name of neuroendocrine tumors (NET) we mean the tumors deriving from the diffuse neuroendocrine cells localized within the gastrointestinal and bronchial epithelium (former known under the name of APUD system cells). The older name of these tumors was "carcinoids". The endocrine pancreatic tumors are also included in the NET category. The vast majority of these tumors express multiple sst receptors. A lack or weak expression of sst receptors occurrs more often in poorly differentiated NET. The data on the incidence of the particular subtypes of sst receptors in NET are rather concordant. The high incidence of sst2 and rare expression of sst 4 was underlined.[13-15] Papotti et al[14] have found that in gastrointestinal NET sst1 and sst2 are the subtypes most commonly detected, whereas sst3 and sst5 are expressed by approximatively two-third of cases. Among the hormonally active pancreatic tumors a preferential expression of sst5 was found in "somatostatinomas", sst2 in "gastrinomas" and "glucagonomas", whereas "insulinomas" express all sst receptor subtypes with prevalence of sst2 and sst5.[14,16] The nonfunctioning tumors express all subtypes except sst4.[14] The membrane localization of sst2 receptor proteins is prevalent, the other subtypes exhibit often a cytoplasmic or mixed (membrane-cytoplasmic) localization (Fig. 1D,E). Our own unpublished preliminary results concern the immunohistochemical detection of sst1-5 in the neuroendocrine tumors from eight patients. We confirm the high incidence of sst1 (6/8), sst2A (5/8), sst3 (4/8), sst5 (5/8). In contrast, the immunostaining for sst4 was negative in all specimens.

Thyroid Tumors

Somatostatin receptors are abundantly expressed in parafollicular (C) cells- derived medullary thyroid cancers (MTC). In the large series of MTC examined by means of immunohistochemistry by Papotti et al,[17] 85% expressed at least one sst receptor subtype. Approximatively half of MTC expressed sst1, sst2, sst3 or sst5. The only subtype rarely detected was sst4 (only 4%). In thyrocyte-derived tumors, sst1-5 were not detected by Northern blot analysis in follicular adenomas and papillary cancers but irregularly expressed in Hürthle cell adenomas and Hürthle cells cancers.[18] However, the cited authors detected sst2 subtype in all investigated tumors using another technical approach (ribonuclease protection assay). Thus, in contrast to MTC, the incidence of sst receptors in differentiated thyroid cancers is controversial and needs further studies.

Adrenal Tumors

Normal adrenal cortex and medulla possess somatostatin receptors. The receptors are also expressed in the adrenal tumors, both in that of medullar origin (pheochromocytomas) as well as in adrenocortical tumors. The immunohistochemical study of sst1-5 receptors in pheochromocytomas was published by Mundschenk et al.[19] The quoted authors found the positive staining for sst3 receptor subtype in 90% of tumors. In contrast, sst2A subtype was expressed only in 25% of tumors. The remaining subtypes were less frequently detected. The more recent data[20] confirm the high incidence of sst3 in pheochromocytomas; the immunopositivity for sst1, sst2A and sst5 was revealed in about one third of the investigated tumors. Among the adrenocortical tumors, functioning adenomas manifesting themselves as Conn's syndrome or Cushing's syndrome, all subtypes of sst receptors are expressed. Contrarly, the nonfunctioning cortical adenomas express mainly sst5; sst1, sst2A and sst 3 were found only in one-third of tumors and sst4 was absent. In adrenal carcinomas, about 50% express sst2A and/or sst3.[20] It should be underlined that in immunopositive tumors, both adreno-cortical and medullar (pheochromocytomas), only a part of tumoral cells was immunostained.

Nonendocrine Cancers

Somatostatin receptors are expressed or even over-expressed also in several nonendocrine cancers. They were revealed, among others, in **breast cancer** tumor samples and cell lines. The positive immunostaining with sst2A and/or sst2B antibodies was detected in 85% of breast cancer samples.[21] Even the higher percent of the investigated samples (98%) expressed sst2 mRNA.[22] The remaining subtypes were also identified, either by immunohistochemis-try, or by PCR. The immunopositivity for sst1 were detected in 52% and in 42% for sst3.[21] The presence of mRNA for sst1 was detected in 91% of samples, and for sst3, sst4 and sst5, in 96%, 76% and 54% of samples, respectively.[22] The expression level of sst2 receptors is higher in cancer tissue than in nonneoplastic surrounding tissues.[23] The incidence of sst1, 2 and 4 correlate positively with estrogen receptor levels, and sst2 additionally with progestrone receptors.[22,23] Moreover, high level of sst2 negatively correlated with proliferative potential of the tumor as measured by Ki-67 staining and seems to be a marker of better prognosis.[23] The data on the incidence of sst receptors in **prostate cancers** are controversial. Using the receptor autoradiography, Reubi et al[24] have found the binding of radiolabeled SST-28, but not of octreotide, to the prostate cancer tissue. This finding suggests the presence of sst receptor different from sst2. By means of in situ hybridization and receptor autoradiography, the quoted authors revealed a preferential expression of sst1.[24,25] In turn, Dizeyi et al,[26] using the RT-PCR and immunohistochemical method, demonstrated subtypes sst1, 2 and 3 in both tumoral and nontumoral epithelial cells, whereas sst4 was expressed preferentially in malignant epithelial cells and sst5 was absent. The **lung cancers** mainly express sst2 (68% of samples examined immunohistochemically). The incidence of other subtypes are scarcer.[27] Interestingly, the sst2 expression is not limited to small cell lung cancer (SCLC), which originates from the neuroendocrine cells of bronchial epithelium, but can be revealed also in nonSCLC. Recently, a functional sst2 was found in a nonSCLC cell line, Calu-6.[28] Since melanocytes are known to derive from the neural crest, **melanoma** is also considered as a neuroendocrine cancer. The high incidence of sst receptors was found in these tumors by means of RT-PCR. Ninety-six % of samples expressed sst1, 83% sst2, 61% sst3, 57% sst4 whereas only 9% sst5.[29] The immunohistochemical findings are roughly similar.[30] However, the data on functional state of these receptors are controversial and the recent study leads to a conclusion that sst receptors in melanoma are not functional.[31] The data on somatostatin receptor detection in gastrointestinal cancers are scarce, with exception of neuroendocrine tumors. The latter are well known to express abundantly the somatostatin receptors and are discussed in the paragraph above. In **gastric adenocarcinomas**, the expression of both sst3 mRNA and protein was found in cancer cells.[32] However, the expression was lower than in normal gastric mucosa. In **colorectal adenocarcinomas**, the overexpression of sst5 was no-ticed.[33,34] Conversely, the expression of sst2, sst3 and sst4 is low. In most nonendocrine **pancreatic cancers**, the sst receptors are lacking.[35,36] This lack is probably important for etiopathogenesis, since sst2 gene transfer to pancreatic cancer cells inhibits their growth.[37] In **hepatocellular cancers**, sst1 and sst2 are mostly expressed, with intermediate expression of sst3 and low of sst4 and sst5.[38] However, the detailed studies of the quoted authors did not reveal any antiproliferative or pro-apoptotic effect of selective SST agonists on sst receptors-expressing hepatocellular cancer lines in vitro. Thus, it seems that these receptors are not fucnctional. The next malignant tumor expressing abundantly sst2 subtype receptors is **renal cell cancer**. The incidence of other subtypes was sst1>sst4>sst3; no data on sst5 were reported.[39] The early studies, based on the in vitro receptor autoradiography, suggested the presence of sst receptors (presumably of sst2 subtype) in nonHodgkin and Hodgkin lymphomas.[40] The more recent data confirm the expression of sst2 and sst3 by means of RT-PCR, but not by means of immunohistochemistry.[41]

Clinical Relevance of Somatostatin Receptors Detection

The direct antiproliferative effects of SST analogs as well as the positive effect of sst-receptor-targeted radiotherapy are determined by at least the presence of sst receptors. Some limited therapeutic effect of SST analogs can be achieved also in sst receptors-negative tumors, via growth hormone/insulin-like growth factor inhibition and/or antiangiogenic action. On the other hand, the detection of sst receptors does not predict that they are functional. Several observations reporting the discrepancies between the receptor incidence and therapeutic effects were published.[31,38,42] The above mentioned discrepancies may depend on structural alterations of receptor proteins or on the alteration of the post-receptor events. It seems that the membrane localization of receptor proteins is also necessary for their biological activity, but this question needs further studies. The coexistence of different subtypes of sst receptors could be also relevant, since heterodimerization of sst receptors can change their functional response. Lastly, it is not fully established what is the significance of sst2 splicing variants, sst2A and sst2B. The in vitro experiments done by Alderton et al[43] suggest that these two variants mediate the opposite effects on cell growth (sst2A being responsive for the antiproliferative effect). Such a possibility is supported by our recent finding, that the in vitro effect of selective sst2 agonist correlates with sst2A, but not with sst2B expression.[44]

Conclusions

Human tumors often express or even over-express the somatostatin receptors. These receptors can be detected in vivo, by means of the receptor scintigraphy, and in vitro (ex vivo) in tumor samples obtained by surgery or by biopsy. Although the receptor scintigraphy presents the clear advantages, the ex vivo detection allows the precise determination of the particular receptor subtypes and their precise cellular localization. Among the different in vitro (ex vivo) techniques, the immunohistochemistry is the most suitable for the routine diagnostic application. The pituitary adenomas and so-called neuroendocrine tumors (carcinoids and neuroendocrine pancreatic tumors) represent the tumors particularly abundantly expressing the sst receptors. However, other endocrine and nonendocrine cancers may also express sst receptors, what make them candidates to the therapeutic trials with SST analogs. On the other hand, this possibility is limited by the fact that the detectable sst receptors are not always functional.

References

1. Panetta R, Patel YC. Expression of mRNA for all five human somatostatin receptors (hSSTR1-5) in pituitary tumors. Life Sci 1995; 56:333-342.
2. Greenman Y, Melmed S. Expression of three somatostatin subtypes in pituitary adenomas: Evidence of preferential SSTR5 expression in the mammosomatotroph lineage. J Clin Endocrinol Metab 1994; 79:724-729.
3. Miller GM, Alexander JM, Bikkal HA et al. Somatostatin receptor subtype gene expression in pituitary adenomas. J Clin Endocrinol Metab 1995; 80:1386-1392.
4. Nielsen S, Mellemkjaer S, Rasmussen LM et al. Expression of somatostatin receptors on human pituitary adenomas in vivo and ex vivo. J Endocrinol Invest 2001; 24:430-437.
5. van der Hoek J, Waaijers M, van Koetsveld PM et al. Distinct functional properties of native somatostatin receptor subtype 5 compared with 2 in the regulation ACTH release by corticotroph tumor cells. Am J Physiol Endocrinol Metab 2005; 289:E278-287.

6. Pawlikowski M, Pisarek H, Kunert-Radek J et al. Immunohistochemical detection of somatostatin receptor subtypes in "clinically nonfunctioning" pituitary adenomas. Endocrine Pathol 2003; 14:231-238.
7. Warnet A, Harris AG, Bernard E et al. Prospective multicenter trial of ocreotide in 24 patients with visual defects caused by nonfunctioning and gonadotropin-secreting pituitary adenomas. Neurosurgery 1997; 41:786-797.
8. Saveanu A, Morange-Ramos I, Gunz G et al. A luteinizing hormone, alpha-subunit, and prolactin-secreting pituitary adenoma responsive to somatostatin anlogs: In vivo and in vitro studies. Eur J Endocrinol 2001; 145:35-41.
9. Reubi JC, Maurer R, Klijn JGM et al. High incidence of somatostatin receptors in human meningiomas: Biochemical characterization. J Clin Endocrinol Metab 1986; 63:433-438.
10. Arena S, Barbieri F, Thellung S et al. Expression of somatostatin receptor mRNA in human meningiomas and their implication in in vitro antiproliferative activity. J Neurooncol 2004; 66:155-166.
11. Fruhwald MC, Rickert CH, O'Dorisio MS et al. Somatostatin receptor subtype 2 is expressed by supratentorial primitive neuroectodermal tumors of childhood and can be targeted for somatostatin receptor imaging. Clin Cancer Res 2004; 10:2997-3006.
12. Fruhwald MC, O'Dorisio MS, Pietsch T et al. High expression somatostatin receptor subtype 2 (sst2) in medulloblstoma: Implications for diagnosis and therapy. Pediatr Res 1999; 45:697-708.
13. Hofland LJ, Liu Q, Van Koetsveld PM et al. Immunohistochemical detection of somatostatin receptor subtypes sst1 andsst2A in human somatostatin receptor positive tumors. J Clin Endocrinol Metab 1999; 84:775-780.
14. Papotti M, Bogiovanni M, Volante M et al. Expression of somatostatin receptor types 1-5 in 81 cases of gastrointestinal and pancreatic tumors. Virchows Arch 2002; 440:461-475.
15. Reubi JC. Somatostatin and other peptide receptors as tools for tumor diagnosis and treatment. Neuroendocrinology 2004; 80(suppl. 1):51-56.
16. Bertherat J, Tenebaum F, Perlemoine K et al. Somatostatin receptors 2 and 2 are the major somatostatin receptors in insulinomas: An in vivo and in vitro study. J Clin Endocrinol Metab 2003; 88:5353-5360.
17. Papotti M, Kumar U, Volante M et al. Immunohistochemical detection of somatostatin receptor types 1-5 in medullary carcinoma of the thyroid. Clin Endocrinol (Oxf) 2001; 54:641-649.
18. Forssell-Aronsson EB, Nilsson O, Bejegard SA et al. 111In-DTPA-D-Phe-1-octreotide binding and somatostatin receptor subtypes in thyroid tumors. J Nucl Med 2000; 41:636-642.
19. Mundschenk J, Unger N, Schulz S et al. Somatostatin receptor subtypes in human pheochromocytoma: Subcellular expression pattern and functional relevance for octreotide scintigrapjy. J Clin Endocrinol Metab 2003; 88:5150-5157.
20. Unger N, Serdiuk I, Sheu SY et al. Immunohistochemical determination of somatostatin receptor subtypes 1, 2A, 3, 4 and 5 in various adrenal tumors. Endocr Res 2004; 30:931-934.
21. Schulz S, Schulz S, Schimtt J et al. Immunocytochemical detection of somatostatin receptos sst1, sst2A, sst2B and sst3 in paraffin embedded breast cancer tissue using subtype-specific antibodies. Clin Cancer Res 1998; 4:2042-2052.
22. Kumar U, Grigorakis SI, Watt HL et al. Somatostatin receptors in primary human breast cancer: Quantitative analysis of mRNA for subtypes 1-5 and correlation with receptor protein expression and tumor pathology. Breat Cancer Res Treat 2005; 92:175-186.
23. Orlando C, Raggi CC, Bianchi S et al. Measurement of receptor subtype 2 mRNA in breast cancer and corresponding normal tissue. Endocr Relat Cancer 2004; 11:323-332.
24. Reubi JC, Waser B, Schaer JC. Somatostatin receptors in human prostate and prostate cancer. J Clin Endocrinol Metab 1995; 80:2806-2814.
25. Reubi JC, Waser B, Schaer JC et al. Somatostatin receptors sst1-sst5 expression in normal and neoplastic human tissues using receptor autoradiography with subtype-selective ligands. Eur J Nucl Med 2001; 28:836-846.
26. Dizeyi N, Konrad L, Bjartell A et al. Localization and mRNA expression of somatostatin receptor subtypes in human prostatic tissue and prostate cancer cell lines. Urol Oncol 2002; 7:91-98.
27. Papotti M, Croce S, Bello M et al. Expression of somatostatin receptor types 2, 3, and 5 in biopsies and surgical specimens of human lung tumours. Correlation with preoperative octreotide scintigraphy. Virchows Arch 2001; 4439:787-797.
28. Ferone D, Arvigo M, Semino C et al. Somatostatin and dopamine receptor expression in lung carcinoma cells and effects of chimeric somatostatin-dopamine molecules on cell proliferation. Am J Physiol Endocrinol Metab 2005; 289:E1044-E1050.
29. Lum SS, Flatcher WS, O'Dorisio MS et al. Distribution and functional significance of somatostatin receptors in malignant melanoma. World J Surg 2001; 25:407-412.
30. Ardjomand N, Ardjomand N, Schaffler G et al. Expression of somatostatin receptors in uveal melanomas. Invest Ophtalmol Vis Sci 2003; 44:980-987.

31. Valencek J, Heere-Ress E, Traub-Weidinger T et al. Somatostatin receptor scintigraphy with 111In-DOTA_Tyr3-octreotide in patients with stage IV melanoma: In vitro and in vivo results. Melanoma Res 2005; 15:523-529.
32. Hu C, Yi C, Hao Z et al. The effect of somatostatin and SSTR3 on proliferation and apoptosis of gastric cancer cells. Cancer Biol Ther 2004; 3:726-730.
33. Vauquereaux V, Dutour A, Bourhim N et al. Increased expression of the mRNA encoding the somatostatin receptor subtype five in human colorectal adenocarcinoma. J Mol Endocrinol 2000; 24:397-408.
34. Vauquereaux V, Grino M, Fina F et al. Somatostatin (SS) receptors (hsst) in colorectal cancer. Eur J Endocrinol 2003; 148(suppl. 1):113, (abstr).
35. Fisher WE, Muscarella P, O'Dorisio TM et al. Expression of the somatostatin receptor subtype-2 gene predicts the response of human pabcreatic cancer to octreotide. Surgery 1996; 120:234-241.
36. Li M, Fisher WE, Kim HJ et al. Somatostatin, somatostatin receptors, and pancreatic cancer. World J Surg 2005; 29:293-296.
37. Du ZY, Quin RY, Xia W et al. Gene transfer of somatostain receptor type 2 by intratumoral injection inhibits established pancreatic carcinoma xenografts. World J Gastroenterol 2005; 11:516-520.
38. Reynaert H, Rombouts K, Vandermonde A et al. Expression of somatostatin receptors in normal and cirrhotic human liver and hepatocellular carcinoma. Gut 2004; 53:1180-1189.
39. Vikie-Topie S, Raisch KP, Kvols LK et al. Expression of somatostatin receptor subtypes in breast carcinoma, carcinoid tumor, and renal cell carcinoma. J Clin Endocrinol Metab 1995; 80:2974-2979.
40. Reubi JC, Waser B, van Hagen M et al. In vitro and in vivo detection of somatostatin receptors in human malignant lymphomas. Int J Cancer 1992; 50:895-900.
41. Dalm VA, Hofland LJ, Mooy CM et al. Somatostatin receptors in malignant lymphomas: Targets for radiotherapy? J Nucl Med 2004; 45:8-16.
42. Plockinger U, Reichel M, Fett U et al. Preoperative octreotide tretment of growth hormone-secreting and clically nonfunctioning pituitary acroadenomas: Effect on tumor volume and lack of corretlation with immunohistochemistry and somatostatin receptor scintigraphy. J Clin Endocrinol Metab 1994; 79:1469-1423.
43. Alderton F, Fan TP, Schindler M. Rat somatostatin sst2(a) and sst2(b) receptor isoforms mediate opposite effects on cell proliferation. Br J Pharmacol 1998; 125:1630-1633.
44. Gruszka A, Kunert-Radek J, Radek A et al. The effect of selective sst1, sst2, sst5 somatostatin receptor agonists, a somatostatin/dopamine chimera and bromocriptine on the "clinically nonfunctioning" pituitary adenomas in vitro. Life Sci 2006; 78:689-693.

CHAPTER 5

The Treatment of Acromegaly with Somatostatin Analogs

Nuria Sucunza, Mª José Barahona and Susan M. Webb*

Abstract

Although transsphenoidal surgery of pituitary micro- or macroadenomas is the treatment of choice in most acromegalic patients, somatostatin analogs are the first choice for medical treatment, either primary or secondary. Nowadays, two forms of octreotide, s.c. short- and i.m. long-acting, and two depot preparations of lanreotide, i.m. slow-repeatable and s.c.autogel, are available. Suppression of GH hypersecretion, lowering IGF-I production, and symptom control, including headache, soft tissue swelling, arthralgia, carpal tunnel syndrome, snoring, sweating and fatigue, are established benefits of the therapy. Metabolic (insulin resistance, hypertriglyceridemia and hypertension), respiratory and cardiac profiles may also improve during somatostatin analogs treatment. In addition, clinically significant tumor shrinkage has been shown in a number of studies, especially in naive patients although the different radiological techniques and criteria of tumor shrinkage used by the authors are not always comparable. Pain at the injection site, cholesterol gallstone development, abdominal pain, diarrhea, fat malabsorption, nausea and flatulence are the main side effects of these drugs. This review summarizes current knowledge of the effects of somatostatin analogs in the treatment of acromegaly.

Introduction

Acromegaly is a rare disease usually caused by a benign pituitary somatotroph adenoma. The disorder usually develops over many years due to a long term exposure to elevated levels of GH. The annual incidence of acromegaly is approximately 3-4 cases/million and its prevalence was estimated at 40-60 cases/million population,[1] and more recently as much as 120.[2] Due to its insidious nature, the diagnosis may be considerably delayed, by 4 to 10 or more years.[1]

Thranssphenoidal adenomectomy remains first-line treatment in most cases, with follow up treatment consisting of medical therapy and/or radiotherapy. Somatostatin (SMS) analogs, available since the mid-1980s, play an important role in the treatment of residual disease (secondary treatment) and as a primary therapy, before surgery or in patients not eligible for surgery. Although symptom relief and tumor shrinkage are obtained in many cases treated with SMS analogs, inadequate suppression of GH secretion is encountered in 40-50% of cases.[3] The first SMS analog available was octreotide with a plasma half-life of approximately 90 minutes; it inhibited GH more potently than insulin or glucagon. Three long- acting forms of SMS analogs, octreotide LAR (long-acting repeatable), lanreotide SR (slow release) and lanreotide autogel are now available. While octreotide is the only analog currently available for clinical use in the United States, in Europe, lanreotide is available as well.[4,5]

*Corresponding Author: Susan M. Webb—Department of Endocrinology, Hospital Sant Pau. Pare Claret 167, 08025-Barcelona. Spain. Email: swebb@santpau.es

Somatostatin Analogs in Diagnostics and Therapy, edited by Marek Pawlikowski. ©2007 Landes Bioscience.

Treatment Algorithm of Acromegaly

Effective treatment is aimed at improving survival and ameliorating many of the deleterious signs and symptoms of acromegalic patients. Survival has been strongly correlated with reduced post-treatment GH levels.[6] Thus, the primary goal of treatment for acromegaly is to normalize GH (≤1 µg/L within 2 hours after 75 oral g glucose load) and serum IGF-I (age and gender matched).[7]

Surgery and Radiotherapy

Surgery remains the first-line treatment in most cases. GH-producing adenoma should be removed completely, with preservation of the pituitary function. This procedure generally results in a rapid and substantial reduction of serum GH immediately postoperatively and normalization of IGF-I in the weeks following the surgery,[8] particularly when the patient is operated by an experienced pituitary surgeon. Nevertheless, only 70-80% of patients with microadenomas and less than 50% with macroadenomas achieve a circulating GH level less than 5 µg/L after surgery.[8] Pituitary radiotherapy can be used as an adjuvant treatment but it may take 5-10 years to lower GH to acceptable levels and induces hypopituitarism.[9,10]

Somatostatin Analogs

SMS Analog Pharmacokinetics

SMS is an endogenous molecule secreted by hypothalamic neurons; via the hypophyseal portal circulation it exerts a variety of physiological effects, including inhibition of GH secretion.[7] The effects of SMS are mediated by SMS receptors, of which there are five different subtypes. SMS receptors subtypes 2 and 5 are the predominant forms found in the pituitary GH-secreting pituitary adenomas.[11] Due to the short half-life of native SMS (1-3 minutes) and its low selectivity,[12] longer acting and more subtype-selective SMS analogs have been used to treat acromegaly. SMS analogs in clinical use today (octreotide and lanreotide) are relatively specific for the subtype 2 receptor. Long-term efficacy of these drugs is correlated with the number, distribution and activity of the SMS receptors on the tumor cells.[13,14] Tachyphylaxis is uncommon, since there is no down regulation phenomenon or decrease of receptor numbers after chronic ligand stimulation.[15] Octreotide, the first available SMS analog for clinical use, is at least 40 times more potent than the native SMS molecule[16] and has a serum half life of approximately 1.5-2 hours after s.c. injection.[7,17,18] Both depot formulations, octreotide-LAR and lanreotide SR, consist of active compounds encapsulated within microspheres of a biodegradable polylactide-polylycolide copolymer that after i.m. injection result in a biphasic drug release.[5] Lanreotide autogel is a newer long-acting aqueous preparation which is administered by deep s.c. injections and provides a consistent drug release during 4 or more weeks.[18-20]

The maximal suppressive effect of s.c. octreotide on GH levels occurs between 2 and 6 hours after the injection.[21] GH levels rise between injections given every 8 hours. The typical dose of octreotide in its s.c. form is 100-250 µg three times a day, but doses up to 1500 µg daily can be given. After a deep i.m. injection of octreotide-LAR, octreotide levels rise briefly within 1 hour, then fall and then begin to rise again about 7-14 days after the injection and remain elevated for an average of 34 days.[22] Steady-state condition is usually achieved after 2 or 3 injections. The usual starting dose of octreotide-LAR is 20 mg monthly. Adjusting doses (down to 10 or up to 30-40 mg every 28 days) are needed based on the response of GH and IGF-I circulating levels.[17,18] After injection of i.m. lanreotide SR, after an initial peak, lanreotide levels remain elevated for about 11 days[23] which means that 10-14 days interval injections are required. The interval may need to be increased or decreased in some patients depending on the GH circulating levels achieved. The natural drug in lanreotide autogel congeals into a slow release aqueous gel that can be given by deep s.c. injections once every 4 and up to 8 weeks.[18] The pharmacological effects can also be manipulated by varying the dose form 60, to 90 and 120 mg with a fixed monthly or bimonthly administration schedule.[18-20]

Indications of SMS Analogs in Acromegaly

SMS analogs can be used as primary treatment, in patients without prior surgery or irradiation, in selected newly diagnosed cases.[24] This concept has gained popularity in recent years and is based on the idea that acromegalic patients, with little prospect of complete surgical resection of their pituitary adenoma, could benefit from medical treatment.[25] Factors favouring the primary use of SMS analogs include low cure rates achieved after surgical resection for pituitary macroadenomas, especially in the presence of cavernous sinus invasion, wide variability in surgical experience, prevalence of perioperative side effects, unacceptable anesthetic risk for some patients, cardiovascular or pulmonary complications and refusal of surgery.[26-29] Acromegalic complications, including insulin resistance, hypertriglyceridemia, hypertension, sleep apnea syndrome or cardiac indices, improve with primary treatment and may be beneficial on hemodynamic outcome of surgery.[24] Shrinkage of tumor before surgery with SMS analogs treatment has not been proven to be useful in facilitating their complete resection.[24]

SMS analogs play an important role in secondary treatment when surgery did not completely suppress GH hypersecretion or while awaiting radiotherapy to be effective.

Biochemical Control with Somatostatin Analog Therapy

Reports on the efficacy of SMS analogs in both forms, short- and long-acting vary considerably. This may be explained by the different methods of GH and IGF-I measurement and by the differences in the inclusion and outcome criteria of each study. Prior sensitivity to a single s.c. octreotide injection has been used to assess the efficacy of long-acting forms of SMS analogs and select patients before starting therapy, while in other instances, all patients have been treated. Although most patients treated with SMS analogs will have some fall in circulating GH,[30] much fewer have persistent and sufficient hormonal suppression.[17] A number of studies have compared the efficacy of short-acting s.c. octreotide, octreotide-LAR and lanreotide SR and most reported slightly greater efficacy with octreotide-LAR.[17] Newman et al[31] found that GH levels decreased in 50-80% of patients treated with octreotide s.c. three times daily and up to 50% of acromegalic patients were considered controlled, (i.e., fasting GH serum levels <2 µg/L and/or normal IGF-I matched by age and gender). Similar results were obtained with 30 mg of i.m. injected lanreotide SR administered once every 10 or 14 days.[19,31] A multicenter study, published in 2000, performed in Belgium and Italy,[32] found no differences in efficacy when comparing s.c octreotide and i.m. lanreotide SR (in conditions of increasing frequency of lanreotide to once every 10 days in 29% and to once every week in 16% of patients). In a systematic review of 125 patients[33] comparing data published on octreotide-LAR and lanreotide SR, slightly better results with octreotide-LAR were found. Controlled GH levels (<2 µg/L) occurred in a mean of 56% of cases treated with octreotide-LAR and in 49% with lanreotide SR, while and IGF-I normalization occurred in 66% and 48% of patients respectively[17] (see Fig. 1).

A metanalysis recently published included 44 trials of over 3 months duration with octreotide-LAR or lanreoride up to 2003 (excluding autogel), with clearly reported data on biochemical efficacy and/or tumor shrinkage;[4] control was defined as normalisation of IGF-I and GH <2.5 µg/L. Twelve studies including 612 patients could evaluate secondary octreotide-LAR, with a mean age around 50 years and a 4 week dose interval in 98.5% of cases; in 69.3% (424/612) prestudy octreotide responsiveness was used as an inclusion criteria and 94% had prior treatment (surgery, radiotherapy, s.c. octreotide or lanreotide); 11 studies were open label, and 1 randomised controlled and blind. For secondary treatment with lanreotide 19 studies including 914 patients, aged a mean of 50 years, of which 31% (283/914) fulfilled the prestudy octreotide responsiveness inclusion criteria were reported; mean duration was also 15 months and 89% had prior treatment (surgery, radiotherapy, or octreotide) and all were open label studies. Doses ranged from 30 to 60 mg and injection intervals ranged from 7 to over 21 days (most frequently every 14 days).

When primary vs. secondary SMS analog therapy was compared, mean GH/IGF-I did not differ after primary or secondary octreotide-LAR therapy, but IGF-I normalisation occurred

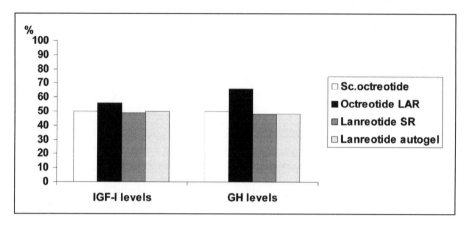

Figure 1. Percentage of patient that achieved GH ≤2 µg/L and normalized IGF-1 levels with secondary treatment with different SMS analogs. Source: Developed from data reported in reference 17.

more after secondary octreotide-LAR than after primary treatment. When pretherapy GH and IGF-I were correlated in relation to efficacy, overall, basal GH was a negative predictor of biochemical cure (p = 0.05); for secondary therapy GH was significant for lanreotide (p <0.05) but not for octreotide-LAR, and for primary therapy basal GH was not significant. As expected, mean pretherapy GH was greater in primary than in secondary octreotide-LAR, but IGF-I did not predict efficacy. Their final conclusions were that octreotide-LAR was more effective than lanreotide in unselected patients in inhibiting GH and IGF-I, that preselection was a positive predictor of IGF-I normalisation and tumor shrinkage, which was greatest with octreotide-LAR; even though biochemical efficacy was similar, shrinkage was greater after primary than after secondary therapy. Fewer studies compare lanreotide autogel and octreotide LAR and no differences in hormonal control or in tolerability have been reported, even though local reactions appear to be less after the autogel.[34-36] In clinical practice, patient's compliance with depot forms will result in better control with these formulations;[37] for convenience patients prefer on the whole less frequent injection requirements. Another advantage with lanreotide autogel is that selected patients may be capable of self-administering the drug, a convenient advantage, since it reduces the patient's dependence on the health system.

SMS Analogs in Combination Therapy

SMS analog therapy in acromegaly can be given in combination with the dopamine agonists bromocriptine or cabergoline. The efficacy of these combinations has been examined in a few small studies. Most have shown that 10-20% of SMS analog resistant patients exhibit some further suppression of GH and/or IGF-I levels with the addition of a dopamine agonist.[17,38,39] Hyperprolactinemia does not seem to be necessary to predict a response to dopamine agonists.[39,40]

Pegvisomant, a biosynthetic analog of human GH that acts as a GH receptor antagonist, is effective in the treatment of acromegaly.[17,28,41] Trainer et al[41] reported a significant and dose-dependent reduction in IGF-I levels, of 26.7%, 50.1% and 62.5%, with 10, 15 or 20 mg of daily s.c. pegvisomant at the end of the study, normal age adjusted IGF-I levels were obtained in 54, 81 and 89% of patients treated respectively. GH levels increased dose-dependently.[41] Normalization of serum IGF-I with pegvisomant has been reported in patients who had failed to respond to SMS analogs.[41,42] Cotreatment of SMS analogs with pegvisomant appears to have an additive effect in terms of reducing IGF-I circulating levels, but endogenous serum GH increase to levels above that measured when SMS analogs are administered alone.[3] This rise in GH induced by pegvisomant is suggested to be caused by increased secretion from residual tumor tissue as a consequence of lowered IGF-I levels and blockade of autofeedback inhibition by GH itself.[3] This promising combination therapy for resistant patients reduces the cost of an expensive drug (since pegvisomant can be given weekly instead of daily),[43] but has yet to be approved by the health authorities and should be considered only in research settings. However, at present data available on sustained effects with prolonged pegvisomant treatment which implies that pegvisomant treatment is limited; furthermore, even though in the USA it as been authorised as primary treatment, in Europe it is only licensed as secondary therapy in operated patients in whom GH is not suppressed after SMS analog treatment.

Effects of SMS Analogs on Tumor Size

SMS analogs can reduce pituitary tumor size but different studies are difficult to compare, since reports on changes in tumor size as an outcome of therapy are not always available and those that do, use a variety of methodologies to define tumor shrinkage (absolute or percentage values, diameter or volume changes). Most define a 10-25% reduction in tumor volume as significant tumor shrinkage. Furthermore, computed tomography and magnetic resonance imaging are used in different studies with different resolutions for each technique. Freda[17] focussed on the data obtained with SMS analogs and found an about 30% tumor shrinkage in patients receiving either the long-acting analogs or s.c. octreotide as adjunctive treatment, together with tumor enlargement in less than 1%. On SMS analogs discontinuation, tumor regrowth appears to occur within 6 months.[44]

In the metanalysis published by Freda et al in 2005,[4] tumor shrinkage greater than 10% was shown to, occur in a greater proportion on octreotide-LAR (47%) compared with lanreotide SR secondary treated patients (21%) after a mean of 15 months of therapy; after primary treatment with octreotide LAR tumor shrinkage was significantly more frequent (88.5%) than after secondary treatment (47%, p <0.01), but not different from that observed after lanreotide (50%,). Tumor shrinkage of at least 10% was also more frequent in patients treated with octreotide LAR vs. s.c. octreotide.[4] Increasing number of months of therapy beyond 6 months was not associated with an increased likehood of tumor shrinkage.[4] Overall tumor increase was observed in 1.4%. A further recent study performed in subjects who received any kind of SMS analogs as primary treatment reported 50% decrease in pituitary mass[26] greater than the tumor shrinkage observed in 4-5% of patients treated with dopamine agonists as primary therapy.[45-48] Another systematic review published by Bevan in 2005 reported tumor shrinkage in 45% of patients treated with s.c. octreotide, in 57% patients treated with octreotide-LAR and in 30% of patients treated with lanreotide SR[18] (see Fig. 2). The same study found that in 921 patients treated with any kind of SMS analog, only 20 (2.2%) experienced further tumor enlargement; thus, more than 97% of patients attain control of tumor growth on SMS analog therapy, at least for periods of up to 3 years. This is a much higher percentage than that of patients who will experience biochemical control, suggesting that mechanisms other than those involved in controlling hormonal hypersecretion determine tumor growth control.

Predictors of tumor shrinkage with SMS analog therapy are summarized in Table 1.[44] Tumor shrinkage is greater in naive patients who have not undergone previous surgery or radiotherapy.

Table 1. Reported predictors of tumor shrinkage after SMS analog therapy

- Patients status regarding prior therapy (newly diagnosed patient vs. post surgery or radiotherapy)
- Tumor size
- Biochemical response
- SMS analog dosage

Source: Developed from data reported in reference 44.

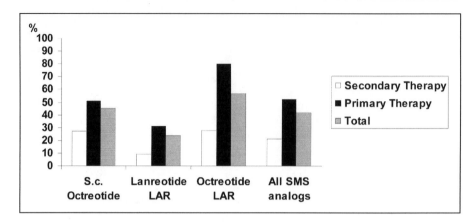

Figure 2. Percentage of patients that achieved tumor shrinkage with SMS analogs. Source: Developed from data reported in reference 18.

Primary therapy reduced mass tumor in a mean of 52% of patients while adjuvant therapy in only 20%.[18,26,44] Tumor shrinkage is greater in macroadenomas than in microadenomas but overall, both can shrink in response to SMS therapy.[44,49-51] While some studies suggest that biochemical response is another predictor of tumor shrinkage[44,52,53] others show no relationship.[49,50,54] The influence of SMS analog dose on tumor mass change is not clear; while Ezzat et al,[44] reported less tumor shrinkage, with 100 µg of s.c. octreotide three times a day than with 250 µg three times a day (19% versus 37%), Cozzi et al[49] found no relationship between tumor shrinkage and octreotide-LAR dose.

Clinical Outcome with Somatostatin Analogs

Clinical improvement of acromegaly occurs overall in about two thirds of patients treated primarily with depot SMS analogs[17] and relatively rapidly. Depot and short acting SMS analogs are similarly effective at controlling signs and symptoms.[17,34,44,55] Improvements in headache, soft tissue swelling, arthralgia, carpal tunnel syndrome, snoring, sweating, fatigue and malaise have been reported[17] (see Table 2). Symptomatic improvement occurs without complete normalization of GH and IGF-I levels. Patients treated with Pegvisomant, that decreases IGF-I and increases GH levels, also have a significant improvement in the signs and symptoms of active acromegaly.[41] Treatment with SMS analogs also reduces glucose,

---

Table 2. Clinical improvement after primary treatment with somatostatin analogs

- Control of headache
- Reduced edema and soft tissues thickness (included tongue volume)
- Reduced arthralgia pain
- Reduced sweating
- Reduced blood pressure
- Reduced insulin resistance and hypertriglyceridemia
- Improvement in severity of sleep apnea
- Reduced left ventricular mass and function, at rest and in response to exercise, and improvement of cardiac failure

Source: Developed from data reported in references 17 and 56-62.

cholesterol and triglycerides levels and improves glycemic and blood pressure control in diabetic and hypertensive acromegalic patients respectively.[56-58]

Cardiovascular disease (left ventricular mass index, ventricular hypertrophy and ejection fraction) also improves with SMS analog treatment,[17,59-61] as well as sleep apnea.[62]

Glucose Homeostasis

Active acromegaly is frequently associated with impaired glucose tolerance or diabetes mellitus,[63-65] mainly caused by insulin resistance.[66-68] Long-term treatments can worsen glucose tolerance, particularly in the postprandial state, because of insulin secretion inhibition.[69] Koop et al[70] showed that 20% of patients treated with octreotide developed impaired glucose tolerance and a further 29% became diabetic; on the other hand, lowering GH levels decreases insulin resistance which contributes to improve carbohydrate metabolism in acromegalic patients.[21,30,71-73]

Baldelli et al showed a significant increase in insulin sensitivity and an improvement of hypertriglyceridemia after initiating SMS analog treatment.[74] The mechanism is suggested to be mediated by the reduction of both GH and glucagon levels, rather than a direct drug effect. Despite the improvement of insulin resistance, a reduced and delayed peak of insulin secretion was observed during an oral glucose tolerance test, which is likely to be a direct effect of SMS analog on β-cell function.[74]

Side Effects

Side effects of the four available forms of SMS analogs are comparable. The primary side effect, which occurs in up to 25% of patients, is increased risk of asymptomatic cholesterol gallstone development, related to the inhibitory effect of SMS analogs on biliary motility and alterations in the bile components.[7,49] Ideally, patients should have a gallbladder ultrasound before initiating SMS analog therapy: the development of symptomatic gallstones should be managed as in sporadic cases.[69] Pain at the injection site is also frequently reported after any of the long-acting SMS analogs (20-30% of patients), while this is less common after s.c.octreotide (7%).[5] Short term side effects, self-limiting within a few weeks without discontinuation of treatment, are abdominal pain, diarrhea, fat malabsorption, nausea and flatulence. Calcium malabsorption can occur with long-term treatments with increased PTH concentration (secondary hyperparathyroidism).[75] Clinically insignificant bradycardia occurs in approximately 25% on patients.[44,71] Pain at the injection site and gastrointestinal symptoms were the main causes of discontinuation of treatment.[34] In a large multicenter study performed in France, 7% of patients treated with lanreotide SR, stopped the treatment because of adverse events.[76]

Conclusion

SMS analogs have been successfully used to treat patients with GH secreting pituitary adenomas; they are safe, effective and usually well tolerated. The results of studies evaluating biochemical, clinical, and tumor size outcome as well as safety and tolerability of the SMS receptor ligands, octreotide and lanreotide, support the use of these agents for both primary and secondary medical therapy in acromegaly.

References

1. Alexander S, Appelton D, Hall R et al. Epidemiology of acromegaly in the Newcastle region. Clin Endocrinol (Oxf) 1980; 12(1):71-9.
2. Kauppinen-Mäkenin R, Sane T, Reunanen A et al. A natiowide survey of mortality in acromegaly. J Clin Endocrinol Metab 2005; 90:4081-6.
3. Jorgensen JO, Feldt-Rasmussen U, Frystyk J et al. Cotreatment of acromegaly with a somatostatin analog and a growth hormone receptor antagonist. J Clin Endocrinol Metab 2005; 90:5627-32.
4. Freda PU, Katznelson L, Van del Lely AJ et al. Long acting somatostatin analog therapy of acromegaly: A metanalisis. J Clin Endocinol Metab 2005; 90:4465-73.

5. Chanson P. Somatostatin analogs in the treatment of acromegaly: The choice is now possible. Eur J Endocrinol 2000; 143(5):573-5.
6. Rajasoorya C, Holdaway IM, Wrightson P et al. Determinants of clinical outcome and survival in acromegaly. Clin Endocrinol (oxf) 1994; 41:95-102.
7. Melmed S, Jackson I, Kleinberg D et al. Current treatment guidelines for acromegaly. J Clin Endocinol Metab 1998; 83:2646-52.
8. Fahlbusch R, Honegger J, Schott W et al. Results of surgery in acromegaly. In: Wass JAH, ed. Treating Acromegaly. Bristol: Bioscientoifica, 1994:49-54.
9. Powell JS, Wardlaw SL, Post KD et al. Outcome of radiotherapy for acromegaly using normalization of insulin-like growth factor I to define cure. J Clin Endocrinol Metab 2000; 85(5):2068-71.
10. Epaminonda P, Porretti S, Cappiello V et al. Efficacy of radiotherapy in normalizing serum IGF-I, acid-labile subunit (ALS) and IGFBP-3 levels in acromegaly. Clin Endocrinol (Oxf) 2001; 55(2):183-9.
11. Reubi JC, Kvols L, Krenning E et al. Distribution of somatostatin receptors in normal and tumor tissue. Metabolism 1990; 39(9 Suppl 2):78-81.
12. Albareda MM, Webb SM. Acromegalia: Tratamiento. Endocrinología 1998; 45(4)156-62.
13. Reubi JC, Landolt AM. The growth hormone responses to octreotide in acromegaly correlate with adenoma somatostatin receptor status. J Clin Endocrinolol Metab 1989; 68:844-50.
14. Ikuyama S, Nawata H, Kato K et al. Plasma growth hormone responses to somatostatin (SRIH) and SRIH receptors in pituitary adenomas in acromegalic patients. J Clin Endocrinol Metab 1986; 62:729-33.
15. Hukovik N, Panetta R, Kumar V et al. Agonist-dependent regulation of cloned human somatostatin receptor type 1-5: Subtype selective internalization or up-regulation. Endocrinology 1996; 137:4046-9.
16. Bauer W, Briner U, Doepfner W et al. SMS 201-995: A very potent and selective optapeptide analog of somatostatin with prolonged action. Life Sci 1982; 31:1133-40.
17. Freda PU. Clinical review 150: Somatostatin analogs in acromegaly. J Clin Endocrinol Metab 2002; 87(7):3013-8.
18. Bevan JS. The antitumoral effects of somatostatin analog therapy in acromegaly. J Clin Endocrinol Metab 2005; 90:1856-63.
19. Caron P, Bex M, Cullen DR et al. One-year follow-up of patients with acromegaly treated with fixed or titrated doses of lanreotide Autogel. Clin Endocrinol (Oxf) 2004; 60(6):734-40.
20. Caron P, Beckers A, Cullen DR et al. Efficacy of the new long-acting formulation of lanreotide (lanreotide Autogel) in the management of acromegaly. J Clin Endocrinol Metab 2002; 87(1):99-104.
21. Ho KY, Weissberger AJ, Marbach P et al. Therapeutic efficacy of the somatostatin analog SMS 201-995 (octreotide) in acromegaly: Effects of dose and frequency and long-term safety. Ann Intern Med 1990; 112:173-81.
22. Stewart PM, Kane KF, Stewart SE et al. Depot long-acting somatostatin analog (sandostatin LAR) is an effective treatment for acromegaly. J Clin Endocrinol Metab 1995; 45:3267-72.
23. Heron I, Thomas F, Dero M et al. Pharmacokinetics and afficacy of a long.acting formulation of the new somatostating analog BIM 23014 in patients with acromegaly. J Clin Endocrinol Metab 1993; 76:721-7.
24. Anat BS, Melmed S. The role of pharmacotherapy in perioperative management of patients with acromegaly. J Clin Endocrinol Metab 2003; 88(3):963-8.
25. Sheppard MC. Primary medical therapy for acromegaly. Clin Endocrinol (Oxf) 2003; 58(4):387-99.
26. Melmed S, Sternberg R, Cook D et al. A critical analysis of pituitary tumor shrinkage during primary medical therapy in acromegaly. J Clin Endocrinol Metab 2005; 90(7):4405-10.
27. Del Pozo C, Webb SM, Oliver B et al. Tratamiento de la acromegalia. Resultados en 56 pacientes. Med Clin (Barc) 1990; 94:85-7.
28. Melmed S, Casanueva FF, Cavagnini F et al. Guidelines for acromegaly management. J Clin Endocrinol Metab 2002; 87(9):4054-8.
29. Clemmons DR, Chihara K, Freda PU et al. Optimizing control of acromegaly: Integrating a growth hormone receptor antagonist into the treatment algorithm. J Clin Endocrinol Metab 2003; 88(10):4759-67.
30. Lamberts SWJ, Oosterom R, Neufeld M et al. The somatostatin analog SMS 201-995 induces long acting inhibition of growth hormone secretion without rebound hypersecretion in acromegalic patients. J Clin Endocrinol Metab 1985; 60:1161-5.
31. Newman CB. Medical therapy for acromegaly. Endocrinol Metab Clin North Am 1999; 28(1):171-90.
32. Verhelst JA, Pedroncelli AM, Abs R et al. Slow-release lanreotide in the treatment of acromegaly: A study in 66 patients. Eur J Endocrinol 2000; 143(5):577-84.

33. McKeage K, Cheer S, Wagstaff AJ. Octreotide long-acting release (LAR): A review of its use in the management of acromegaly. Pituitary 2003; 63(22):2473-99.

34. Alexopoulou O, Abrams P, Verhlest J et al. Efficacy and tolerability of lanreotide Autogel therapy in acromegalic patients previously treated with octreotide LAR. Eur J Endocrinol 2004; 151(3):317-24.

35. Ashwell SG, Bevan JS, Edwards OM et al. The efficacy and safety of lanreotide Autogel in patients with acromegaly previously treated with octreotide LAR. Eur J Endocrinol 2004; 150(4):473-80.

36. Van Thiel SW, Romijn JA, Biermasz NR et al. Octreotide long-acting repeatable and lanreotide Autogel are equally effective in controlling growth hormone secretion in acromegalic patients. Eur J Endocrinol 2004; 150(4):489-95.

37. Baldelli R, Colao A, Razzore O et al. Two-year follow up of acromegalic patients treated wit slow release lanreotide (30mg). J Clin Endocrinol Metab 2000; 85:4099-103.

38. Lamberts SW, Zweens M, Verschoor L et al. A comparison among the growth hormone-lowering effects in acromegaly of the somatostatin analog SMS 201-995, bromocriptine, and the combination of both drugs. J Clin Endocrinol Metab 1986; 63(1):16-9.

39. Selvarajah D, Webster J, Ross R et al. Effectiveness of adding dopamine agonist therapy to long-acting somatostatin analogues in the management of acromegaly. Eur J Endocrinol 2005; 152(4):569-74.

40. Cozzi R, Attanasio R, Lodrini S et al. Cabergoline addition to depot somatostatin analogues in resistant acromegalic patients: Efficacy and lack of predictive value of prolactin status. Clin Endocrinol (Oxf) 2004; 61(2):209-15.

41. Trainer PJ, Drake WM, Katznelson L et al. Treatment of acromegaly with the growth hormone-receptor antagonist pegvisomant. N Engl J Med 2000; 342(16):1171-7.

42. Van der Lely AJ, Hutson RK, Trainer PJ et al. Long-term treatment of acromegaly with pegvisomant, a growth hormone receptor antagonist. Lancet 2001; 358(9295):1754-9.

43. Feenstra J, de Herder WW, ten Have SMT et al. Combined therpay with somatostatin analogues and weekly pegvisomant in active acromegaly. Lancet 2005; 365:1644-6.

44. Ezzat S, Snyder PJ, Young WF et al. Octreotide treatment of acromegaly: A randomized, multicenter study. Ann Intern Med 1992; 117:711-8.

45. Colao A, Ferone D, Marzullo P et al. Effect of different dopaminergic agents in the treatment of acromegaly. J Clin Endocrinol Metab 1997; 82(2):518-23.

46. Sachdev Y, Gomez-Pan A, Tunbridge WM et al. Bromocriptine therapy in acromegaly. Lancet 1975; 2(7946):1164-8.

47. Oppizzi G, Liuzzi A, Chiodini P et al. Dopaminergic treatment of acromegaly: Different effects on hormone secretion and tumor size. J Clin Endocrinol Metab 1984; 58(6):988-92.

48. Abs R, Verhelst J, Maiter D et al. Cabergoline in the treatment of acromegaly: A study in 64 patients. J Clin Endocrinol Metab 1998; 83(2):374-8.

49. Cozzi R, Attanasio R, Montini M et al. Four-year treatment with octreotide-long-acting repeatable in 110 acromegalic patients: Predictive value of short-term results? J Clin Endocrinol Metab 2003; 88(7):3090-8.

50. Bevan JS, Atkin SL, Atkinson AB et al. Primary medical therapy for acromegaly: An open, prospective, multicenter study of the effects of subcutaneous and intramuscular slow-release octreotide on growth hormone, insulin-like growth factor-I, and tumor size. J Clin Endocrinol Metab 2002; 87(10):4554-63.

51. Amato G, Mazziotti G, Rotondi M et al. Long-term effects of lanreotide SR and octreotide LAR on tumour shrinkage and GH hypersecretion in patients with previously untreated acromegaly. Clin Endocrinol (Oxf) 2002; 56(1):65-71.

52. Abe T, Ludecke DK. Effects of preoperative octreotide treatment on different subtypes of 90 GH-secreting pituitary adenomas and outcome in one surgical centre. Eur J Endocrinol 2001; 145(2):137-45.

53. Arosio M, Macchelli S, Rossi CM et al. Effects of treatment with octreotide in acromegalic patients—A multicenter Italian study. Italian Multicenter Octreotide Study Group. Eur J Endocrinol 1995; 133(4):430-9.

54. Cozzi R, Barausse M, Sberna M et al. Lanreotide 60 mg, a longer-acting somatostatin analog: Tumor shrinkage and hormonal normalization in acromegaly. Pituitary 2000; 3(4):231-8.

55. Vance ML, Harris AG. Long-term treatment of 189 acromegalic patients with the somatostatin analog octreotide. Results of the International Multicenter Acromegaly Study Group. Arch Intern Med 1991; 151(8):1573-8.

56. Colao A, Ferone D, Cappabianca P et al. Effect of octreotide pretreatment on surgical outcome in acromegaly. J Clin Endocrinol Metab 1997; 82:3308-14.

57. Stevenaert A, Beckers A. Presurgical octreotide: Treatment in acromegaly. Metabolism 1996; 45(8 Suppl 1):72-4.
58. Fahlbusch R, Honegger J, Buchfelder M. Acromegaly—The place of the neurosurgeon. Metabolism 1996; 45(8 Suppl 1):65-6.
59. Colao A, Cuocolo A, Marzullo P et al. Effects on 1 year treatment with octreotide on cardiac performance in patients with acromegaly. J Clin Endocrinol Metab 1999; 84:17-23.
60. Baldelli R, Ferretti E, Jaffrani-Rea ML et al. Cardiac effects on the slow release lanreotide, a slow release somatostatin analog, in acromegalic patients. J Clin Endocrinol Metab 1999; 84:625-32.
61. Hradec J, Oral J, Janota T et al. Regression of acromegalic left ventricular hypertrophy after lanreotide (a show release somatostatin analog, in acromegalic patients. Am J Cardiol 1999; 83:1506-9.
62. Grunstein RR, Ho KK, Sullivan CE. Effect of octreotide, a somatostatin analog, on sleep apnea in patients with acromegaly. Ann Intern Med 1994; 121(7):478-83.
63. Mestrón A, Webb SM, Astorga R et al. Epidemiology, clinical characteristics, outcome, morbidity and mortality in acromegaly based on the Spanish Acromegaly Registry (Registro Español de Acromegalia, REA). Eur J Endocrinol 2004; 151(4):439-46.
64. Holdaway IM, Rajasoorya C. Epidemiology of acromegaly. Pituitary 1999; 2(1):29-41.
65. Colao A, Ferone D, Marzullo P et al. Systemic complications of acromegaly: Epidemiology, pathogenesis, and management. Endocr Rev 2004; 25(1):102-52.
66. Colao A, Lombardi G. Growth-hormone and prolactin excess. Lancet 1998; 352(9138):1455-61.
67. Melmed S. Acromegaly. N Engl J Med 1990; 322(14):966-77.
68. Baldelli R, Diéguez C, Casanueva FF. Acromegaly: From basic research to clinical studies. Polish J Endocrinol 2001; 52:57-79.
69. Van der Lely AJ, Beckers A, Daly AF et al. Acromegaly: Pathology, diagnosis and treatment. In: Taylor, Francis, eds. Somatostatin Analogs. 2005:109-20.
70. Koop BL, Harris AG, Ezzat S. Effect of octreotide on glucose tolerance in acromegaly. Eur J Endocrinol 1994; 130(6):581-6.
71. Newman B, Melmed S, Snyder PJ et al. Safety and efficacy of long term octreotide therapy of acromegaly: Results of a multicenter trial in 103 patients-a clinical research center study. J Clin Endocrinol Metab 1995; 80:2768-75.
72. Lamberts SWJ, Reubi JC, Krenning EP et al. Somatostatin analogs in the treatment of acromegaly. Endocrinol Metab Clin North Am 1992; 21:737-52.
73. Lamberts SW, Uitterlinden P, Verschoor L et al. Long-term treatment of acromegaly with the somatostatin analogue SMS 201-995. N Engl J Med 1985; 313(25):1576-80.
74. Baldelli R, Battista C, Leonetti F et al. Glucose homeostasis in acromegaly: Effects of long-acting somatostatin analogues treatment. Clin Endocrinol (Oxf) 2003; 59(4):492-9.
75. Cappelli C, Gandossi E, Agosti B et al. Long-term treatment of acromegaly with lanreotide: Evidence of increased serum parathormone concentration. Endocr J 2004; 51(6):517-20.
76. Chanson P, Leselbaum A, Blumberg J et al. Efficacy and tolerability of the long-acting somatostatin analog lanreotide in acromegaly. A 12-month multicenter study of 58 acromegalic patients. French Multicenter Study Group on Lanreotide in Acromegaly. Pituitary 2000; 2(4):269-76.

CHAPTER 6

The Treatment of Neuroendocrine Tumors (NET) with Somatostatin Analogs

Beata Kos-Kudla*

Abstract

There is increasing interest in somatostatin (SST) analogs in the diagnosis and therapy of neuroendocrine tumors (NET), which constitute a heterogeneous group of neoplasms often associated with typical symptoms due to excessive production of hormones and bioactive substances. SST analogs have been proven as drugs of first choice in the reliable control of hormone-mediated symptoms and for nonresectable gastroenteropancreatic (GEP) NET. Somatostatin can exert both cytotoxic and cytostatic actions. Evidence for antiproliferative properties of SST and its analogs derives from in vitro and in vivo studies. SST analogs show also antitumor activity, however tumor shrinkage in GEP NET has rarely been observed, but tumor stabilization in 25-75% of patients with NET has been reported.

The high expression of SST receptors in endocrine tumors has provided the molecular basis for the successful use of radiolabelled SST analogs as tumor tracers in nuclear medicine for diagnosis and radiotherapy. This is a useful palliative option for symptomatic patients with inoperable or metastatic tumors.

Further prospective randomized studies are needed to assess the exact role of SST analogs in the treatment of NET. Many questions still have to be solved in clinical trials. Recently intensive research is also focused on the development of new SST analogs. In the future, they will probably improve the diagnosis and management of neuroendocrine tumors.

Introduction

Neuroendocrine tumors (NET) constitute a heterogeneous group of neoplasms that are often associated with typical symptoms due to excessive and uncontrolled production of hormones and bioactive substances. The choice of treatment for NET depends primarily on the pathologic differentiation and stage at diagnosis but also on the presence of symptoms related to hormonal secretion. Often, these tumors are slow growing and sometimes can be managed with clinical observation only. However, they can display accelerated progression, requiring a much more aggressive strategy. Surgery remains the first line therapy in NET. However, these tumors are usually diagnosed at an advanced stage when cure cannot be achieved. Somatostatin (SST) analogs are important agents in the medical treatment of gastroenteropancreatic (GEP) NET. They only can not reduce secretion of hormones and hence control hormone-related symptoms but also are capable of controlling disease progression.[1]

*Beata Kos-Kudla—Department of Pathophysiology and Endocrinology, Silesian Medical Academy, Pl.Traugutta 2, 41-800 Zabrze, Poland. Email: beatakos@ka.onet.pl

Somatostatin Analogs in Diagnostics and Therapy, edited by Marek Pawlikowski.
©2007 Landes Bioscience.

Somatostatin is a natural small cyclic peptide hormone synthesized as part of a large prohormone molecule that is enzymatically cleaved into its active form.[1] (see also Chapter 1 of this book). SST is found in classic "open type" endocrine cells from which it is directly excreted into the blood, paracrine cells with long cytoplasmic extensions which terminate on putative effector cells and in neurones where it may function as a neurotransmitter or is released from the nerve endings into the blood. Thus, depending on its site of elaboration, SST may function as a hormone, a neurohormone, a neurotransmitter or a parahormone.[2]

The longer-acting analogs octreotide and lanreotide are modifications of the naturally occurring somatostatin-14. Size reduction and the amino acid modifications protect the molecule against enzymatic degradation and prolong the plasma half-life of octreotide to approximately 1,5 h and of lanreotide to 2,5 h vs. the 2-3 min half-life of somatostatin-14.[3] Somatostatin activity is mediated through five specific SST receptors (sst) (sstr-1 to sstr-5) located on the membrane of the target cells (see also Chapter 2). Expression of sst2 and sst5 is particularly high in pancreatic NET (including gastrinomas, glucagonomas, and VIPomas) and gut NETs (foregut, midgut and hindgut).[4] Both sstr-2 and sstr-5 are found in approximately 90% and 80% of tumors, respectively, making these tumors potentially sensitive to hormonal treatment that targets these receptors.[1,5] Insulinomas have a lower incidence (50-70%).[4] Octreotide acetate and lanreotide bind preferentially to sst2 and to a lower degree to sstr-5.[3] Undifferentiated NET express somatostatin receptors less often (and in lower density) than well-differentiated ones. The somatostatin receptor distribution is generally homogeneous in NET. Among the 5 somatostatin receptor subtypes, the sstr-2 subtype is the most frequently expressed, while sstr 4 is rarely detected. sstr-2 is predominantly membrane bound.[4] A recent review reported that the expression of sstr-1-5 in GEP NET was 68, 86, 46, 93 and 57% respectively.[6]

Individual GEP NET express multiple receptor subtypes (see also Chapter 5). The role and relevance of these receptors in tumor growth and symptom control has not been fully elucidated. Different tumors may express assorted subtypes, the density of which may vary in different areas. It is postulated that actions of SST analogs mediated by the sstr-2 undergo tachyphylaxis, which leads to a loss of response in many carcinoid patients.[7]

Somatostatin and its long-acting analogs have been proven as drugs of first choice in the reliable control of hormone - mediated symptoms.[3,8] SST acts on various targets including the brain, pituitary, pancreas, gut, adrenals, thyroid, kidney and on the immune system to regulate a variety of physiological functions. Its actions include inhibition of endocrine and exocrine secretions, modulation of neurotransmission, motor and cognitive functions, inhibition of intestinal motility, absorption of nutrients and ions, vascular contractility and cell proliferation.[9]

Prolonged-release formulations now allow drug administration every 2 to 4 weeks. Both octreotide and lanreotide have been shown to be efficacious in managing symptoms and tumor progression compared with standard doses of short-acting SST analogs.[1,10]

Lanreotide is a long-acting somatostatin analog that requires intramuscular injection two or three times per month and has also been shown to be effective both in controlling symptoms and in reducing tumor markers in patients with carcinoid tumors.[11-13] The treatment of patients with carcinoid syndrome with lanreotide would be expected to achieve greater acceptability to patients and to improve quality of life. In the O'Toole and coworkers[13] study the clinical and biologic efficacy of octreotide and lanreotide in a crossover analysis of patients with symptomatic carcinoid syndrome was compared and both treatment modalities in terms of acceptability to patients and treatment preference were assessed. Although no differences in quality-of-life scores were observed between those who received octreotide and those who received lanreotide, patients significantly preferred lanreotide over octreotide largely due to the simplified mode of administration.[13] Treatments with both somatostatin analogs were remarkably well tolerated.

The long-acting somatostatin analogs are effective in controlling symptomatic carcinoid syndrome and in lowering biochemical tumor markers. Lanreotide's and octreotide's efficacy in controlling the debilitating symptoms associated with carcinoid syndrome makes them invaluable for a large proportion of patients who suffer from this disease.[13]

Control of Symptoms and Biochemical Responses in Functionally Active GEP NET

Octreotide controlled symptoms caused by hormone overproduction in about 70-90% of carcinoid patients. In a direct comparison, diarrhea and flushing were successfully controlled with octreotide in 50 and 68% of carcinoid patients, respectively. Similar potencies on diarrhea and flushing (45 and 54%, respectively) have been reported with lanreotide in patients with functional GEP NET. Although symptoms improve initially in most patients, a loss of response occurs in about 50% with continuous treatment.[7,13,14] In parallel to a reduction in carcinoid symptoms, a reduced excretion of the serotonin metabolite, 5-hydroxyindoleacetic acid (5-HIAA), in urine as well as the matrix protein of large dense core vesicles , chromogranin A, in serum is observed during somatostatin analog therapy. Side-effects leading to termination of treatment are rare and are only to some extent dose-dependent.[15]

In the first trial reported by Kvols et al.,[16] octreotide subcutaneously (150 μg t.i.d.) was observed to present symptomatic responses in 88% and biochemical responses in 72% of patients with carcinoid tumors. The median duration of the biochemical response was 12 months. In 1989 Gorden et al[17] performed a meta-analysis of all reported cases of neuroendocrine tumors treated with somatostatin analogs. The meta-analysis indicated symptomatic improvement in 92% and a biochemical response in 66% of the patients.[18] More than 50 patients with gastrinomas were treated with doses of 100-1,500 μg octreotide/day, most of them in the short term. A clinical response defined as control of gastric hypersecretion, pain and diarrhea was observed in 90% of the patients and was accompanied by a significant reduction in serum gastrin and basal acid secretion.[18,19] The two most common clinical syndromes related to endocrine pancreatic tumors are the Zollinger-Ellison syndrome resulting from gastrin overproduction and the hypoglycemic syndrome which is related to high insulin/proinsulin release. The Zollinger-Ellison syndrome (gastrinoma) may also arise from hypersecretion of gastrin production by duodenal NET (around 40%). Other functioning endocrine pancreatic tumors are the VIP-producing syndrome or WDHA syndrome which is characterized by extensive diarrhea, hypokalemia and achlorhydria. In such patients the diarrhea volume might exceed more than 10 liters per day and they often require intensive care. Another rare tumor is the glucagonoma. The glucagonoma syndrome is characterized by a typical necrolytic migratory erythema, diabetic glucose tolerance, anemia, weight loss and tromboembolism which may be related because of glucagon production and its end-organ effects. Both these rare tumors have been treated successfully with somatostatin analogs.[18]

SST analog therapy is not the primary treatment for gastrinomas; however, for many patients being initially treated with H_2 receptor blockers or proton pump inhibitors combined with surgery, long-acting SST analogs may be beneficial for a subgroup of patients with malignant gastrinomas. Patients with insulin-producing tumors treated with somatostatin analogs should be very carefully monitored for escalation of their hypoglycemia. About 50% of insulin-producing tumors do not express somatostatin receptor 2 and 5 subtypes. However, there is a subgroup of insulin-producing tumors that might benefit from SST analog therapy and these are the predominantly malignant insulinomas which concomitantly hypersecrete gastrin and/or glucagon. Symptomatic improvement has been reported in more than 80% of patients with WDHA syndrome and VIP-producing tumors treated with octreotide at doses of 100-400 μg/day. Biochemical responses have been reported in about 80% of VIPoma patients. However, in some patients the beneficial effect of octreotide lasted only a few days requiring progressive increases in doses.[18,19]

The combination of somatostatin analogs with other agents in such situations is still an interesting area for future studies. Janson et al[20] have previously reported that a combination of octreotide and interferon-α produced biochemical responses in 75% of patients resistant to either interferon-α alone or conventional doses of octreotide.[18,20]

Carcinoid Syndrome

The occurrence and severity of the syndrome depend on the site of origin of the tumor and tumor mass, but the presence of carcinoid syndrome almost always implies distant metastases in the liver. The syndrome is present in approximately 10% of patients with carcinoid tumors.[21] Somatostatin analogs are the best therapy for controlling the symptoms of carcinoid syndrome.

Long-acting somatostatin analogs such as octreotide acetate and lanreotide have been repeatedly shown to improve or even normalize diarrhea and flushing in patients with carcinoid syndrome during both short- and long-term treatment. Symptom relief is correlated with reduction of hormonal markers. Somatostatin analogs act by inhibition of hormone release from the tumor and by direct inhibition of water and electrolyte secretion from the intestine.[3]

Different formulations of SST analogs are available and all seem to be safe and effective. Octreotide acetate given as subcutaneous injection up to three times daily, intramuscular lanreotide injection given once per 1-2 weeks and monthly intramuscular Sandostatin LAR have demonstrated similar efficacy in short-term studies.[22] Garland et al[22] assessed the long-term effect of long-acting octreotide acetate (Sandostatin LAR) on the management of patients with malignant carcinoid syndrome in the 3-year retrospective study. Sandostatin LAR provides good long-term symptomatic control in these patients; it is well tolerated and patients expressed a preference for monthly intramuscular Sandostatin LAR as opposed to daily subcutaneous injections of octreotide. It improved satisfaction in their management.

More recently Ruszniewski et al[21] showed in a 6-month, open, non-controlled, multicenter study that 28-day prolonged-release (PR) formulation of lanreotide was effective in reducing flushing and diarrhea and biochemical markers associated with carcinoid tumor. The degree of improvement and the good tolerability are consistent with previous studies of other lanreotide formulations. The convenient presentation of the PR formulation in a prefilled syringe facilitates administration, may improve acceptability and should be considered a suitable treatment for patients with carcinoid syndrome.

Antiproliferative Effect of SST Analogs on GEP NET

Evidence for antiproliferative properties of somatostatin and its analogs derives from in vitro and in vivo studies. Somatostatin can exert both cytotoxic and cytostatic actions. It has been demonstrated that it mediated arrest at the G1 phase of the cell cycle and that this effect is mediated through activation of sstr-2 and sstr-5.[3,23,24]

The antiproliferative effects of SST analogs on GEP NET may be the result of direct and indirect actions (Fig. 1).[33]

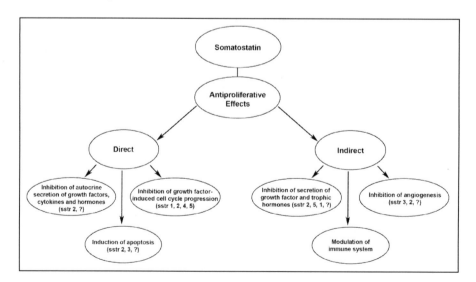

Figure 1. Mechanisms of direct and indirect effects of somatostatin.

Faiss et al[15] showed that ultra-high-dose lanreotide treatment in patients with metastatic GEP NET, progressive under previous therapies (e.g., low-dose somatostatin analogs and/or interferon-α), can generate an additional antiproliferative effect especially in patients with midgut tumor disease. Thus, dose escalation of somatostatin analogs up to ultra-high doses as used in the study represents a useful therapeutic adjuvant in metastatic neuroendocrine tumor disease following conventional therapy.

In contrast to foregut and midgut tumors, they[15] did not observe an antiproliferative effect in hindgut tumors. This confirms the long-stadning hypothesis that metastatic hindgut tumors are refractory to all forms of medical therapies including somatostatin analogs despite the expression of somatostatin receptors in vivo. This again raises the issue that, in addition to the expression of somatostatin receptors in neuroendocrine tumor cells, other humoral and/or cellular components must be involved.[15]

Further studies are needed to assess the exact role of SST analogs in induction of apoptosis and control of growth in GEP NETs. Currently, no convincing evidence exists to support the use of SST analogs at high doses.

Direct Effects of SST on Cell Growth

Somatostatin and its analogs may directly inhibit tumor cell growth by interacting with specific somatostatin receptors located on tumor cells.

Analyses of sstr mRNAs demonstrate that various human tumors from neuroendocrine and gastroenteropancreatic origin express various sstr mRNA, each tumor expressing more than one subtype, sstr-2 being the most frequently expressed.[9,25]

The presence of somatostatin receptors in tumors argues in favor of a direct role for somatostatin in the regulation of tumor growth. A direct inhibitory effect of somatostatin or analogs on cell growth has been demonstrated on various cancer cell lines which express endogenous somatostatin receptors (cells of mammary, pancreatic, gastric, lung, colorectal or thyroid origin). However, the mechanisms of cell growth arrest induced by SST are still poorly understood.[9,26]

The direct inhibitory action of SST on cell growth may result from the blockade of mitogenic growth factor signal. The antiproliferative effect of somatostatin can also result from apoptosis. Apoptosis has been reported to be induced by the sstr-3 subtype via a G protein-dependent signaling and to be associated with an intracellular acidification and activation of endonuclease and induction of p53 and Bax.[9,27,28]

Imam et al[29] evaluated apoptotic effects of octreotide on neuroendocrine tumors using BON-1, a human serotonin-secreting pancreatic endocrine tumor cell line xenografted into nude mice. They showed on increase in apoptotic cells in mice receiving treatment with high dose octreotide compared with a placebo group. In parallel to their preclinical study, Imam et al[29] also studied apoptosis of tissue samples in patients with neuroendocrine tumors treated with high doses of lanreotide. After 6 and 12 months of treatment, patients receiving high-dose SST analog showed a biochemical response (decrease in different neuroendocrine tumor markers) and an increase in apoptotic index (AI). Therefore the treatment with high-dose SST analogs may induce apoptosis in NET, while this was not found during treatment with low-dose SST analogs.[29]

Besides the antiproliferative effect of somatostatin due to cell growth arrest and apoptosis, SST may directly control cell growth by inhibiting the synthesis and the secretion of autocrine growth factors, cytokines and hormones involved in the proliferation of tumor cells. It is well known that the aberrant expression of growth factors, cytokines or hormones and their receptors represent fundamental circuits that may induce uncontrolled growth and metastatic behavior of cancer cells.[9]

Indirect Effects of SST on Cell Growth

The indirect antineoplastic effects of somatostatin and its analogs are independent of the presence of SST receptors on the tumor cells. Therefore, a possible beneficial effect of somatostatin analog therapy is not restricted to tumors expressing SST receptors.[2]

Indirect effects of somatostatin on tumor cell growth may be the result of inhibition of secrection of growth-promoting hormones and growth factors which stimulate the growth of various types of cancer. It is known that tumors depend on specific growth factors for their growth.[9] Insulin-like growth factor-1 (IGF-1) is an important modulator of many neoplasms that express IGF-1 receptors. Octreotide has been demonstrated to negatively control IGF-1 level. In addition, a direct effect on IGF gene expression has also been reported.[9,30]

Somatostatin and its analogs can also indirectly control tumor development and metastasis by inhibition of angiogenesis. They inhibit angiogenesis in vitro and in vivo.[9,31] Somatostatin and its receptors may play a regulatory role in hemodynamic tumor-host interactions and vascular drainage.[9]

Evidence suggests that SST may influence the immune system. Somatostatin and receptors are expressed in human lymphoid organs and can regulate various immune functions including lymphocyte proliferation, immunoglobulin synthesis, and cytokine production. It has been well demonstrated that SS and its analogs inhibit the proliferation of human and rat lympho-cytes in vitro. However, there is no information available on the effect of somatostatin in vivo on the immune system.[9,32]

Control of Tumor Growth in GEP NET

SST analogs show antitumor activity in a variety of experimental models both in vivo and in vitro.[33,34] However, tumor shrinkage in GEP NET has rarely been observed, but SST analogs have been reported to induce tumor volume stabilisation in 25-75% of patients with neuroendocrine tumors.[15,35-42] Faiss et al[15] studied a group of 30 patients with metastatic GEP NET who underwent ultra-high-dose (15 mg/d) lanreotide therapy and evaluated tumor growth every 3 months. After a 1-year treatment period one complete and one partial remission were observed in patients with functional midgut NET. Eleven patients had stable disease and 11 patients showed continuing tumor growth.[15] Shojamanesh et al[35] showed that SS analogs have antitumor growth effects in patients with malignant pancreatic NET. Fifteen patients with malignant gastrinoma were treated initially with octreotide, 200 µg every 12 hours, and at last follow-up were being maintained on long-acting octreotide, 20-30 mg every month. 47% of studied patients demonstrated tumor stabilization and 6% a de-crease in tumor size. The growth response was long-lasting, the mean duration of response was 25.0+/-6.1 months.[35]

Control of tumor growth occurs in approximately 50% of VIPoma patients treated with octreotide. However, the reduction in the size of liver metastases secondary to VIPomas is temporary and transient and control of tumor growth in response to octreotide therapy is lost after 8-16 months of treatment.[2]

Aparicio et al[43] confirmed an antitumor activity of SS analog treatment resulted in 57% of cases of tumor stabilisation in 35 patients with progressive NET. They (ref. 43) stated that somatostatin analogs produce an inhibitory effect that leads to tumor stabilisation in most patients with "slowly progressing" NET. The authors (ref. 43) suggested determining the pre-treatment slope of the tumor growth rate (STGR) as a useful yardstick for selecting patients likely to benefit from SST analog therapy. NET differentiation, the primary tumor site, the carcinoid syndrome and the intensity of octreoscan uptake are also potential predictive factors for response to somatostatin analog therapy.[43]

Eriksson et al[37] reported one major response and disease stabilisation in 70% of 19 cases of GEP NETs over a median duration of 12 months. Nine of the patients in this study had previously been treated with standard doses of octreotide suggesting a possible benefit of dose escalation. Moreover, no enhanced antitumor effect was gained by administering higher doses of SST analogs. The dose-response effect is still poorly understood. The authors[37] stated that no conclusion can be drawn given the small number of patients. Variations in tumor stabilization rates between studies could be explained by the great heterogeneity of NET patients (Table 1).

Table 1. Antitumor activity of somatostatin analogs activity of somatostatin analogs in patients with NETs

	No of Patients	Drug	Tumors Regression	Tumors Stabilisation	Median Duration of Response [Months]
Salz et al[36] 1993	34	250 µg octreotide 3x daily	0%	50%	5
Arnold[38] 1996	52	200 µg octreotide 3x daily	0%	37%	18
Eriksson[37] 1997	19	750 – 12000 µg/day lancreotide	5%	70%	12
Faiss[15] 1999	30	5 mg lancreotide 3x daily	3 %	37 %	
Ducreux[40] 2000	39	30 mg/every 14 days	5%	49%	9,5
Aparicio[43] 2001	35	100 µg octreotide 3x daily 30 mg lancreotide every 14 days or both SS analogs	3 %	57% (76%-slowly 33%-rapidly progressing NET)	11
Faiss[39] 2003	25	1 mg lancreotide 3x daily;	4%	25%	12
		1 mg lancreotide 3x daily+interferon-α 5 x 10^6 IU 3 times a week	7.1%	28%	12
Welin[44] 2004	12	Octreotide pamoate 160 mg every 2 weeks		75%	12

Recently, Faiss et al[39] showed in the prospective, randomized, multicenter study for the first time that somatostatin analogs, interferon alpha, or the combination of the two had comparable antiproliferative effects in the treatment of metastatic neuroendocrine gastroenteropancreatic tumors. Response rates were lower compared with those published in previous, nonrandomized studies. The antiproliferative effect of the tested substances was similar for functional and nonfunctional neuroendocrine tumors.

Most of the studies are comparable in the sense that patients with progressive disease were included. Partial tumor regression was a rare event and stabilization of tumor growth the most favourable result occurring in patients with GEP NET. However, even stable disease is relatively short-lasting since growth inhibition was reported to last for 2-60 months.[3]

Influence of SST Analogs on Survival of Patients with GEP NET

There is, as yet, no published study that shows that therapy with somatostatin analogs, as the only treatment, improves survival.[18] The overall five year survival for pancreatic NETs is 50-80%, with insulinoma and gastrinoma having up to 94% five-year survival, although clearly there is large variation depending on the stage at presentation and whether curative surgery is possible.[8] The 5-year survival rates of patients with neuroendocrine tumors is less than 20% when liver metastases are present. Furthermore, the median survival for patients with malignant carcinoid tumors with the carcinoid syndrome is less than 2 years from the time of diagnosis.[18]

The long-term treatment of patients with carcinoid syndrome with octreotide has resulted in an increase of median progression-free survival from 3.5 to 15 months. The duration of the remission can be short (median 8-12 months), and the early development of tachyphylaxis is not uncommon.[45]

The majority of centers working with patients with neuroendocrine tumors use a multimodal therapeutic approach. Thus, it is very unlikely that a patient with a neuroendocrine tumor will receive a somatostatin analog as the only treatment during the clinical course of the disease.[18]

To date, no prospective randomized trial has been performed in patients with GEP NETs treated with SST analogs compared with observation. Additional studies are necessary to determine any convincing effect of SST analogs on survival.[1]

Adverse Reactions to SST Analog Therapy

The most common side-effects of somatostatin analog are generally mild and include nausea, transient abdominal cramps, flatulence, diarrhea and local reaction at the injection site. Most of these minor side-effects resolve with time. In 2-50% of patients gall stones are formed de novo, but these remain virtually always asymptomatic. Rare, more severe adverse events of octreotide therapy include hypocalcemia, bradycardia, acute pancreatitis, hepatitis, jaundice, transitory ischemic attacks, and a negative inotropic effect of the analogs.[18,46]

Tachyphylaxis and Resistance to SST Analogs

Desensitization to the effect of SST analogs within weeks or months has been observed in the majority of GEP NET-bearing patients, on important difference existing between patients with respect to the induction of the tumor type-dependent tachyphylaxis phenomenon.[33,47]

Potential mechanisms of tachyphylaxis and resistance to somatostatin analog therapy in patients with sst receptor-positive tumors are: receptor down-regulation, desensitization (a decrease in responsiveness due to receptor uncoupling from second messenger activation), nonhomogeneous expression of sstr in tumors, outgrowth of sstr-negative cell clones, resistance due to tachyphylaxis of the inhibitory effect of SST analogs on indirect tumor growth-promoting mechanisms and mutations in sstr genes leading to the absence of functional receptor proteins.[46,47]

Tumor-Targeted Radioactive SST Analog Treatment

The high expression of sstr in endocrine tumors has provided the molecular basis for the successful use of radiolabelled octreotide or lanreotide analogs as tumor tracers in nuclear medicine for diagnosis and radiotherapy (see also Chapter 7). This is a useful palliative option for symptomatic patients with inoperable or metastatic tumor. The principle of treatment is only to give radionuclide therapy when there is abnormally increased uptake of the corresponding imaging agent. No randomised controlled trials have yet been performed.[8,33]

Peptide receptors scintigraphy with a radioactive somatostatin analog (^{111}In-DTPA-octreotide) is a sensitive and specific technique to identify in vivo the presence and abundance of somatostatin receptors in various tumors. The method has now been accepted as an important tool for the staging and localization of neuroendocrine tumors. However, the technique is currently being evaluated as a possible means of targeting neuroendocrine tumors with "tumor-targeted" radiotherapy using a repeated administration of high doses of ^{111}In-DTPA-octreotide.[18] Therapy with ^{111}In-pentetreotide in the progressive metastatic NET (performed in three centres, n = 81 patients) resulted in stable disease in 57% of patients.[49] The ability of sstr receptors to internalize initiated the development of sst receptor-targeted radiotherapy with radiolabelled somatostatin analogs using the β-emitter yttrium- 90 [^{90}Y-DOTA^0,Tyr^3]octreotide] (OctreoTher), [^{90}Y-DOTA]-lanreotide and lutenium177 [^{177}Lu-DOTA^0,Tyr^3]octreotate.[33]

Therapy with somatostatin analogs coupled to β-emitting radionuclides, such as ^{90}Y and ^{177}Lu, is potentially more effective, as higher tumor radiation doses can be achieved and the longer range of the β-particles (1-10 mm) may also lead to radiation of neighbouring receptor-negative tumor cells.[49]

Although there is a need for more accurate prediction of the absorbed doses, treatment with all ^{111}In-, ^{90}Y- or ^{177}Lu-labelled SS analogs has been associated with considerable symptomatic improvement with limited side effects that are mainly related to kidney function.[50] (See also Chapter 7).

New SST Analogs and Future Developments

SOM230 is a new somatostatin analog with high binding affinity to human somatostatin 1,2,3 and sstr-5 receptors. Based on the multiligand binding properties of SOM230, on its prolonged inhibition of hormone secretion in animal models and on the expression of multiple SST receptors in GEP tumors, more patients with GEP NET could be treated successfully with SOM230 without experiencing tachyphylaxis during chronic therapy.[7,51] SOM230 had no agonist activity at the sstr-4, in line with lack of affinity for this subtype. This SST analog was particularly potent at sstr-5 compared with octreotide.[7] As SOM230 strongly inhibits cAMP production by stimulating sstr-1 and sstr-5 and multiple SST receptors are expressed in GEP NET, SOM230 could have a greater inhibitory effect than octreotide on hormones secreted by carcinoid tumors. Moreover, SOM230 may be the only pharmacological treatment for GEP tumor patients who are not responsive to currently available therapy, either because they do not express sstr-2 or because targeting these receptors does not translate into clinical efficacy.[7,52,53]

SST analogs have also been used as carriers to deliver cytotoxic agents to cancer cells. Therapy studies with radiolabelled somatostatin analogs linked with cytotoxic compounds have so far been carried out in experimental tumor models only but they are very promising.[33,49] Schally et al[54] synthesised new targeted cytotoxic somatostatin octapeptide conjugates such as RC-121 and RC-160 coupled to doxorubicin or its superactive derivative 2-pyrrolino-DOX (AN-201). AN-238, which contains AN-201 linked to the carrier RC-121, has been demonstrated to be very effective on a variety of human experimental cancer models.

Recently, a new approach with a genetic radioisotope targeting strategy has been proposed, based on the induction of sst2 receptor expression and the selective tumor uptake of radiolabelled peptides.[55] Of particular importance in the future is the need to exploit the major advances in gene therapy to optimize the full potential of somatostatin analogs in the management of neoplasia.[2]

The combination of somatostatin analogs with interferons and cytotoxic agents in randomized clinical trials is also worthy of investigation.[18] Combination therapy of somatostatin analogs with cytotoxics or other hormonal treatments targeted somatostatin analog chemotherapy or radiotherapy in both advanced malignancy and in the adjuvant setting may prove to be very much more effective than somatostatin analog monotherapy. Carefully controlled clinical studies with objective outcome measures are required to evaluate combination therapies.[2]

Recently intensive research is also focused on the development of new peptidic and non-peptidic somatostatin analogs, selective agonists for each receptor subtype and pansomatostatin analogs with a binding profile similar to that of the natural peptide.

These compounds might help in the characterisation of unsuspected biological activities and new indications for somatostatin. In the future, they will probably improve the diagnosis and treatment of neuroendocrine tumors.[33,56]

References

1. Delaunoit T, Rubin J, Neczyporenko F et al. Somatostatin analogs in the treatment of gastroenteropancreatic neuroendocrine tumors. Mayo Clin Proc 2005; 80(4):502-6.
2. Jenkins SA, Kynaston HG, Davies N et al. Somatostatin analogs in oncology: A look to the future. Chemotherapy 2001; 47(suppl 2):162-196.
3. Arnold R, Simon B, Wied M. Treatment of neuroendocrine GEP tumors with somatostatin analogs. Digestion 2000; 62:84-91.
4. Reubi JC. Somatostatin and other peptide receptors as tools for tumor diagnosis and treatment. Neuroendocrinology 2004; 80(suppl 1):51-56.
5. Wulbrand U, Wied M, Zofel P et al. Growth factor receptor expression in human gastroenteropancreatic euroendocrine tumors. Eur J Clin Invest 1998; 28:1038-1049.

6. Oberg K, Kvols L, Caplin M et al. Consensus report on the use of somatostatin analogs for the gastroenteropancreatic system. Ann Oncol 2004; 15:966-973.
7. Schmid HA, Philippe Schoeffter P. Functional activity of the multiligand analog SOM230 at human recombinant somatostatin receptor subtypes supports its usefulness in neuroendocrine tumors. Neuroendocrinology 2004; 80(suppl):47-50.
8. Ramage JK, Davies AHG, Ardill J et al. Guidelines for the management of gastroenteropancreatic neuroendocrine (including carcinoid) tumors. Gut 2005; 54:iv1-iv16.
9. Bousquet C, Puente E, Buscail L et al. Antiproliferative effect of somatostatin and analogs. Chemotherapy 2001; 47(Suppl 2):30-9.
10. Dogliotti L, Tampellini M, Stivanello M et al. The clinical management of neuroendocrine tumors with long-acting repeatable (LAR) octreotide: Comparison with standard subcutaneous octreotide therapy. Ann Oncol 2001; 12(suppl 2):S105-S109.
11. Scherubl H, Wiedenmann B, Riecken EO et al. Treatment of the carcinoid syndrome with a depot formulation of the somatostatin analogue lanreotide. Eur J Cancer 1994; 30A:1590-1.
12. Ruszniewski P, Ducreux M, Chayvialle JA et al. Treatment of the carcinoid syndrome with the long-acting somatostatin analogue lanreotide: A prospective study in 39 patients. Gut 1996; 39(2):279-83.
13. O'Toole D, Ducreux M, Bommelaer G et al. Treatment of carcinoid syndrome: A prospective crossover evaluation of lanreotide versus octreotide in terms of efficacy, patient acceptability, and tolerance. Cancer 2000; 88(4):770-6.
14. Wymenga AN, Eriksson B, Salmela PI et al. Efficacy and safety of prolonged-release lanreotide in patients with gastrointestinal neuroendocrine tumors and hormone-related symptoms. J Clin Oncol 1999; 17:1111.
15. Faiss S, Rath U, Mansmann U et al. Ultra-high-dose lanreotide treatment in patients with metastatic neuroendocrine gastroenteropancreatic tumors. Digestion 1999; 60:469-476.
16. Kvols LK, Moertel CG, O'Connell MJ et al. Treatment of the malignant carcinoid syndrome. Evaluation of a long-acting somatostatin analogue. N Engl J Med 1986; 315:663-666.
17. Gorden O, Comi RJ, Maton PN et al. Somatostatin analogue (SMS 201-995) in treatment of hormone-secreting tumors of the pituitary and gastrointestinal tract and nonneoplastic diseases of the gut. Ann Intern Med 1989; 110:35-50.
18. Öberg K. Established clinical use of octreotide and lanreotide in oncology. Chemotherapy 2001; 47(suppl 2):40-53.
19. Scarpignato C. Somatostatin analogs in the management of endocrine tumors of the pancreas. In: Mignon M, Jensen RT, eds. Endocrine Tumors of the Pancreas. Basel: Karger: 1995:385-414.
20. Janson ET, Ahlström H, Andersson T et al. Octreotide and interferon alpha: A new combination for the treatment of malignant carcinoid tumors. Eur J Cancer 1992; 28A:1647-1650.
21. Ruszniewski P, Ish-Shalom S, Wymenga M et al. Rapid and sustained relief from the symptoms of carcinoid syndrome: Results from an open 6-month study of the 28-day prolonged-release formulation of lanreotide. Neuroendocrinology 2004; 80:244-5.
22. Garland J, Buscombe JR, Bouvier C et al. Sandostatin LAR (long-acting octreotide acetate) for malignant carcinoid syndrome: A 3-year experience. Aliment Pharmacol Ther 2003; 17(3):437-44.
23. Sharma K, Srikant CB. Induction of wild - type p53, BAX, and acidic endonuclease during somatostatin - Signated apoptosis in MCF-7 human breast cancer cells. Int J Cancer 1998; 76:259-266.
24. Cordelier P, Esteve JP, Bousquet C et al. Characterization of the antiproliferative signal mediated by the somatostatin receptor subtype sstr5. Proc Natl Acad Sci USA 1997; 921:580-584.
25. Patel YC. Molecular pharmacology of somatostatin receptor subtypes. J Endocrinol Invest 1997; 20:348-367.
26. Weckbecker G, Raulf F, Stolz B et al. Somatostatin analogs for diagnosis and treatment of cancer. Pharmacol Ther 1993; 60(2):245-264.
27. Sharma K, Patel YC, Srikant CB. Subtype-selective induction of wild type p53 and apoptosis, but not cell cycle arrest, by human somatostatin receptor 3. Mol Endocrinol 1996; 10:1688-96.
28. Sharma K, Srikant CB. G protein coupled receptor signaled apoptosis is associated with activation of cation insensitive acidic endonuclease and intracellular acidification. Biophys Res Commun 1998; 242:134-140.
29. Imam H, Eriksson B, Lukinius A et al. Induction of apoptosis in neuroendocrine tumors of the digestive system during treatment with somatostatin analogs. Acta Oncol 1997; 36:607-614.
30. Serri O, Brazeau P, Kachra Z et al. Octreotide inhibits insulin - like growth factor - I hepatic gene expression in the hypophysectomized rat: Evidence for a direct and indirect mechanism of action. Endocrinology 1992; 130:1816-1821.
31. Woltering EA, Watson JC, Alperin-Lea RC et al. Somatostatin analogs: Angiogenesis inhibitors with novel mechanisms of action. Invest New Drugs 1997; 15(1):77-86.

32. van Hagen PM, Krenning EP, Kwakkeboom DJ et al. Somatostatin and the immune and haematopoetic system; a review. Eur J Clin Invest 1994; 24(2):91-99.
33. Guillermet-Guibert J, Lahlou H, Pyronnet S et al. Somatostatin receptors as tools for diagnosis and therapy: Molecular aspects. Best Practice and Research Clinical Gastroenterology 2005; 19(4):535-551.
34. Schally AV, Comaru-Schally AM, Nagy A et al. Hypothalamic hormones and cancer. Front Neuroendocrinol 2001; 22:248-291.
35. Shojamanesh H, Gibril F, Louie A et al. Prospective study of the antitumor efficacy of long-term octreotide treatment in patients with progressive metastatic gastrinoma. Cancer 2002; 94:331-343.
36. Saltz L, Trochanowski B, Buckley M et al. Octreotide as an antineoplastic agent in the treatment of functional and nonfunctional neuroendocrine tumors. Cancer 1993; 72:244-248.
37. Eriksson B, Renstrup J, Iman H et al. High-dose treatment with lanreotide of patients with advanced neuroendocrine gastrointestinal tumor: Clinical and biological effects. Ann Oncol 1997; 8:1041-1044.
38. Arnold R, Trautmann ME, Creutzfeldt W et al. The German Sandostatin Multicentre Study Group. Somatostatin analogue octreotide and inhibition of tumor growth in metastatic endocrine gastroenteropancreatic tumors. Gut 1996; 38:430-438.
39. Faiss S, Pape UF, Böhmig M et al. Prospective, randomized, multicenter trial on the antiproliferative effect of lanreotide, interferon alpha, and their combination for therapy of metastatic neuroendocrine gastroenteropancreatic tumors-The International Lanreotide and Interferon Alfa Study Group. J Clin Oncol 2003; 21:2689-2696.
40. Ducreux M, Ruszniewski P, Chayvialle JA et al. The antitumoral effect of the long-acting somatostatin analog lanreotide in neuroendocrine tumors. Am J Gastroenterol 2000; 95:3276-3281.
41. Ricci S, Antonuzzo A, Galli L et al. Octreotide acetate long-acting release in patients with metastatic neuroendocrine tumors pretreated with lanreotide. Ann Oncol 2000; 11:1127-1130.
42. Oberg K. Future aspects of somatostatin-receptor-mediated therapy. Neuroendocrinology 2004; 80(supplement 1):57-61.
43. Aparicio T, Ducreux M, Baudin E et al. Antitumor activity of somatostatin analogues in progressive metastatic neuroendocrine tumors. Eur J Cancer 2001; 37:1014-1019.
44. Welin SV, Janson ET, Sundin A et al. High-dose treatment with a long-acting somatostatin analogue in patients with advanced midgut carcinoid tumors. Eur J Endocrin 2004; 151:107-112.
45. de Herder WW, Lamberts SWJ. Gut endocrine tumors. Best Practice and Research Clinical Endocrinology & Metabolism 2004; 18(4):477-495.
46. Trendle MC, Moertel CG, Kvols LK. Incidence and morbidity of cholelithiasis in patients receiving chronic octreotide for metastatic carcinoid and malignant islet cell tumors. Cancer 1997; 79:830-834.
47. Hofland LJ, Lamberts SW. The pathophysiological consequences of somatostatin receptor internalization and resistance. Endocrine Reviews 2003; 24:28-47.
48. de Herder WW, Hofland LJ, van der Lely AJ et al. Somatostatin receptors in gastroenteropancreatic neuroendocrine tumors. Endocr Relat Cancer 2003; 10:451-458.
49. de Herder WW, Krenning EP, van Eijck CHJ et al. Considerations concerning a tailored, individualized therapeutic management of patients with (neuro)endocrine tumors of the gastrointestinal tract and pancreas. Endocr Relat Cancer 2004; 11:19-34.
50. Kaltsas GA, Papadogias D, Makras P et al. Treatment of advanced neuroendocrine tumors with radiolabelled somatostatin analogs. Endocr Relat Cancer 2005; 12(4):683-99.
51. Weckbecker G, Briner U, Lewis I et al. SOM230: A new somatostatin peptidomimetic with potent inhibitory effects on the growth hormone/insulin-like growth factor-I axis in rats, primates, and dogs. Endocrinology 2002; 143(10):4123-30.
52. Hobday TJ, Rubin J, Goldberg R et al. Molecular markers in metastatic gastrointestinal neuroendocrine tumors. Proc Am Soc Clin Oncol 2003; 22:269.
53. de Herder WW, Hofland LJ, van der Lely SWJ. Somatostatin recptors in gastroenteropancreatic neuroendocrine tumors. Endocr Relat Cancer 2003; 10:451-458.
54. Schally AV, Szepeshazi K, Nagy A et al. New approaches to therapy of cancers of the stomach, colon and pancreas based on peptide analogs. Cell Mol Life Sci 2004; 61:1042-1068.
55. Buchsbaum DJ. Imaging and therapy of tumors induced to express somatostatin receptor by gene transfer using radiolabeled peptides and single chain antibody constructs. Semin Nucl Med 2004; 34:32-46.
56. Reubi JC, Waser B. Concomitant expression of several peptide receptors in neuroendocrine tumors: Molecular basis for in vivo multireceptor tumor targeting. Eur J Nucl Med Mol Imaging 2003; 30:781-793.

CHAPTER 7

Radiolabeled Somatostatin Analogs in the Diagnosis and Therapy of Neuroendocrine Tumors

Leszek Królicki,* Jolanta Kunikowska and Marianna I. Bak

Abstract

The proof of the principle, that many tumors are characterized by a high expression of somatostatin receptors, has provided the background for the use of somatostatin analogs in the diagnostic process and in the therapy of these tumors. At present, somatostatin receptor scintigraphy can be adequately used particularly in neuroendocrine tumors. When compared with other techniques, the sensitivity of radionuclide examination is much greater in the visualization of primary tumors and its metastases. This technique is also used for preliminary assessment of the degree of malignancy, therefore it could serve as a background for the treatment strategy and can help to evaluate progress of the disease.

The analogs of somatostatin were introduced in several therapy modalities because it was shown that they can reduce significantly the progress of the disease. The results of somatostatin-targeted radiotherapy suggest that this kind of treatment modality could be a venue for administration of cytostatic agents and radioisotopes. This highly specific treatment is possible on the basis of the principle that somatostatin analog binds with receptor and that the receptor-radioligand complex is internalized into cancer cell.

This review summarizes the present results on the application of radiolabeled somatostatin analogs at the diagnostic and therapeutic level in certain human cancers.

Introduction

The metabolic and biochemical processes leading to the change of the normal cell into the cancer cell are not known in details, as yet. The main role is played by the changes of the genome. When the factor (usually unidentified) changes the genetic code it leads to the metabolic alterations in the cell and, as consequences, to the changes in the function of the respective organ. Several characteristic changes in the phenotype of the cancer cells are used now in the diagnose making process and in the therapy. For example, the increased consumption of glucose by cancer cells leads to the implementation of 18FDG in PET examination; the increased transport of amino acids leads to the use of radio labeled methionine and tyrosine in the diagnose of some cancers. The other characteristic process in the growth of malignant tumors is the change in the regulation of receptor expression, most importantly for the regulatory peptide hormones, described for the first time in 1984.[1] These peptides can be labeled by

*Corresoponding Author: Leszek Królicki—Department of Nuclear Medicine, Medical Academy of Warsaw, ul.Banacha 1a; 02-097 Warszawa, Poland. Email: krolicki@amwaw.edu.pl

Somatostatin Analogs in Diagnostics and Therapy, edited by Marek Pawlikowski. ©2007 Landes Bioscience.

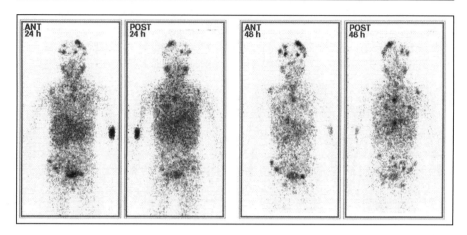

Figure 1. 131I-MIBG scans in a patient with multiple pheochromocytoma metastases.

radionuclide and used in the nuclear medicine procedures. The main feature of these peptides, as a radiopharmaceutics, is the fact that their synthesis is relatively simple and that they can penetrate by diffusion into cancer tumors.

Neuroendocrine tumors (NET) and gastro-entero-pancreatic tumors (GEP) are characterized by an increased uptake of neuroamines and increased expression of somatostatin receptors and therefore they are the main indication for the use of radiopharmaceutics in the diagnosis making process and in therapy. These tumors can be localized by administration of 123I or 131I-meta-iodobenzylguanidine (MIBG) and also by radiolabeled somatostatin analogs. MIBG enters the cell by the active transport (type 1 uptake mechanism) and is used in the diagnostic processes of tumors originating in suprarenal gland, paraganglioma, neuroblastoma and several others.[2,3] Sensitivity of this technique is approximately 90%. 131I-MIBG is also used for the therapy in those tumors (Fig. 1). Somatostatin analogs bind with high affinity to somatostatin receptors which are present in a large variety of NETs, adenocarcinomas, inflammatory and immune cells. Additionally, in some other tumors several radiolabeled peptides different from somatostatin are used in diagnostic and treatment procedures: VIP, gastrin, bombesin, cholecystokinin and substance P.

Implementation of peptides, which have the ability to bind to particular cancer cell receptors, has an impact in several areas of research and practice:

1. chronic medical therapy with use of peptides analogs;
2. diagnostic tests with administration of analogs labeled with radionuclides;
3. therapy with analogs labeled by radionuclides or chemotherapeutics.

Somatostatin and Somatostatin Receptors

Although the expression of particular receptors in the tumor cells is related to various peptides, but only few of them were implemented for the clinical use. Somatostatin analogs were the first peptides widely used in the clinical practice. Somatostatin inhibits a wide spectrum of physiological processes, including peptide hormone secretion. Moreover, somatostatin inhibits several growth factors, angiogenesis. On the other side, it stimulates apoptosis and activates the reticulo-endothelial system and, as a consequence, leads to the inhibition of tumor growth (see also Chapter 1). These anti-tumor effects caused a growing interest in somatostatin during the past two decades.

Somatostatin receptors belong to a superfamily of G protein-coupled receptors that can functionally couple to various intracellular effector systems. Somatostatin receptors are located in the plasma membranes and contain seven transmembrane domains.

The connection of ligand with receptor leads to internationalization of the whole complex through the endocytosis. Inside the cell the ligand is separated and stored in lysosomes.

To date, five human somatostatin receptors subtypes (sst 1-5) have been cloned (see also Chapter 2). Because they have the distinct characteristics they were divided into two classes: the class I includes sst2a, sst3 and sst5 somatostatin receptors; and the class II: sst1 and sst4 receptors. Importantly, each of the somatostatin receptors subtype has a distinct ability of internalization . The final effect of stimulation of somatostatin receptors results from the different effects of all receptor subtypes. It is important to remember that the expression of different subtypes of somatostatin receptors in the different types of NET cancers varies; even in the same type of tumor the expression can change.[4] In most of tumors located in the gut the expression of sst1 and sst2 was found, and only in few of them sst5 was observed. Receptor subtype sst3 was very rarely observed in the gut carcinoids. Receptor sst4 is present in 22-86% of NET tumors. Somatostatin receptors are expressed with various densities in the following tumors: small cell lung cancer, breast, prostate, colon, renal, ovarian, hepatocellular, thyroid cancer, Hodgkin and nonHodgkin lymphomas, and brain tumors. Somatostatin receptors had been detected also in: activated lymphocytes, sarcoidosis, tuberculosis, rheumatoid diseases, Crohn's disease, coeliakia, Hashimoto and Graves' diseases, Wagener's granulomatosis, aspergillosis, lupus erhytematodes, Henoch-Schonlein purpura, and even in haemangiomas.[5-8]

The Analogs of Somatostatin as Radiopharmaceutics

Because the time of disappearance of somatostatin is short (1-2 minutes), it was necessary to find the somatostatin analogs with much prolonged half-life time in plasma and tissues. The use of analogs has several advantages: they have the fast clearance, the fast penetration into the tumors, they do not cause allergic reactions and they are quite simply synthesized and modified. However, in contrary to the native somatostatin, they have the different binding affinities for the particular subtypes of somatostatin receptors. Each analog has its own characteristic and the decision which analog should be used is crucial for the scintigraphy. In the diagnostic tests analogs with the high binding affinity for somatostatin receptors sst2, sst3 and sst5 are used.

Somatostatin analogs as radiopharmaceutics are composed from the three parts: (1) the particular peptide (somatostatin analog), (2) radionuclide, and (3) chelating substance which is the connecting part between somatostatin analog and radioisotope. Each of these elements has its distinct role and everyone is capable to change the pharmacokinetics of radiopharmaceutics and could influence the success of radionuclide image as a diagnostic tool.

At present, three somatostatin analogs were developed for clinical use: octreotide, lanreotide and vapreotide (RC-160). They are all able to bind to sst2 and sst5; vapreotide, in addition, can bind to sst4. The use of iodo-radioisotopes does not need chelator for labeling the peptide; iodine connects directly with amino acids building the peptide molecule. The use of the metallic radioisotopes (111In, 90Y, 99mTc, 177 Lu) needs certain types of chelators, such as DTPA. More advanced chelator molecule is DOTA, which has a specific high affinity for the binding with somatostatin. Recently, HYNIC together with coligand EDDA.is frequently used.[9-13]

From the commercially available radioisotopes used for the labeling of somatostatin analogs the very important is 111In. 111In has a prolonged half-life (2.8 days), therefore scan could be made up to 72 hrs. where the tumor-to-background ratio was increased. However 111In is a radioisotope obtained from cyclotrons, therefore is expensive. Additionally, its energy of radiation is not optimal for typical gamma cameras. Many obstacles related to the use of 111In have been deleted after the introduction of 99mTc; the analogs of somatostatin may be labeled immediately before the diagnostic procedure and 99mTc has a good characteristic of radiation for gamma cameras. currently used.

The first agent which was used in the radioisotope diagnostic test was octapeptide **[Tyr 3] octreotide**, labeled with 123I.[14,15] It has main affinity for sst2, much smaller for sst5 and the smallest for sst3. However, this agent has very fast hepatic clearance resulting in accumulation of radioactivity in liver, gall bladder, bile ducts and in the gastrointestinal tract. These negative factors caused difficulties in the interpretation of scans from the abdomen. The introduction of novel somatostatin analogs and of DTPA and DOTA chelators, was a great improvement in the nuclear medicine because their molecules, eliminated by kidneys, were able to circumvent the disadvantages mentioned before. The first somatostatin analog commercially used which has a chelator in its structure was **111In DTPA-[d-Phe 1]** octreotide (Octreoscan). This agent got a FDA approval in year 1994 for the diagnostic tests of neuroendocrine tumors.

Depreotide has affinity for the sst2, sst3 and sst5 receptors. This analog labeled with 99mTc binds to sst receptors in NET tumors, melanoma, small cells- and nonsmall cells pulmonary tumors. In addition, it has been shown that sst receptors are also present in the tumor itself and in peritumoral tissue, where angiogenesis takes place.

Recently, the new analog was obtained: DOTATATE, which has a similar structure as DOTA-[D-Phe1,Tyr3] octreotide. It has much higher affinity to sst2. It could be labeled with: 111In, 99mTc, 90Y or 177Lu. Table 1 summarizes most frequently used somatostatin analogs in nuclear medicine.

A successful clinical applications of particular sst analogs relies primarily on the presence of tumoral receptors able to bind the peptide analog, and furthermore the next step which is internalization of the ligand-receptor complex. The internalized ligand-receptor complex could be identified as hot spots on gamma-camera scans. In the experiments with rats, where [111In-DTPA]-octreotide was used, has been shown that injection of cold analog caused the inhibition of radiopharmaceutics accumulation in the pituitary gland and pancreas; but analog injection 20 min after radiopharmaceutics did not influence the accumulation. It was speculated that internalization of the ligand-receptor complex takes 20 minutes.[16]

The various subtypes of sst are internalized with different speed. In the study with Chinese hamster ovary cells (CHO)-K1 it has been shown that this process is time and temperature dependent.[17] Rocheville et al[18] demonstrated that internalization of human sst subtypes can be determined by functional homo- and heterodimerization of sst subtypes as well.The internalization of the ligand-receptor complex was observed in vivo in human tumors and also in vitro in removed tumors. Janson et al[19] studied the specimens of carcinoid tissue from the group of 7 patients 2 days after injection of [111In-DTPA] octreotide. In all seven cases scintigraphy made before surgery showed a high accumulation of radiopharmaceutics in the region of tumor localization. The accumulation of sst analog in plasma membrane, cytoplasm, in the secretory granules, vesicular compartments and perinuclear area was shown. In human tumors sst2 receptor is responsible for internalization of [111In-DTPA] octreotide.[20] However, the other subtypes of sst could be also involved in this process.[17] The other studies had shown that complex with [90Y-DOTA0,Tyr3]-octreotide also is internalized and this analog has higher internalization in pancreatic tumors compared with [111In-DOTA, Tyr3]-octreotide and [111In-DTPA0]-octreotide (1.8 fold and 3.5 fold, respectively).[21] In summary, the degree of analogs internalization is as follows:

Hynic 99mTc octreotide = 111In DTPA TOC < 99mTcHynic TOC < 99mTc Tyr-Thr Tate. = 111In Tate.

After internalization of the receptor-radioligand complex, an important process is the retention of radioactivity within the tumor cells.In vitro observations suggest, that about 50% of the internalized radioactivity is released from the cells within 6 hrs, mainly as a radioligand degradation products. However, recycling of the receptor-radioligand complex may play a role as well. Such a process has been demonstrated for sst2 and sst3 receptors.[22] It suggests that trapping of radioisotopes into tumor cells may be an additional mechanism determining the amount of uptake of radioligand that is used for sst receptor scintigraphy or receptor-targeted radiotherapy. Determination of the optimal peptide mass for uptake of radioactivity in human tumors after injection of the radiolabeled SST analogs are ongoing.

The accumulation of the radiolabeled somatostatin analogs is related to the previous treatment with "cold" analogs. This treatment can make up or down regulation of the sst receptors.

Table 1. Somatostatin analogs most frequently used in the nuclear medicine

[111In-DTPA] octreotide
[90Y-DOTA, Tyr^3]octreotide (DOTATOC)
[90Y-DOTA] lanreotide (DOTALAN)
[90Y-DOTA,Tyr^3] octreotate (DOTATATE)
[90Y-DOTA, 1-Nal^3]octreotide
[9mTc-EDDA/HYNIC]Tyr(3)-octreotide
[99Tc]-P829
[99mTc]-depreotide

Table 2. Factors responsible for somatostatin analogs accumulation in cancer tumors

- Biological stability of the radioligand
- Distribution and density of sst receptors expressed on tumors
- Type of sst receptors with increased expression by tumors
- Radioligand affinity for existing sst receptors in tumor cells
- Receptors efficiency for internalization of radioligand
- Radioisotope trapping within the tumor intracellular structures
- Molecular weight of the injected radipharmaceutical
- Past treatment with cold somatostatin analogs

The observations by Ronga et al[23] showed a decreased expression of sst receptors in patients with the thyroid medullary cancer. Dorr et al[24] also showed a decreased expression of sst receptors in liver cells, spleen and kidneys after the treatment with "cold" somatostatin analogs; in the same work an increased affinity to somatostatin receptors in carcinoid cells was noted.

In summary, from all studies published up today,[10] we have to point up that the successful application of the particular radiopharmaceutics in clinical practice is strongly related to the structure of sst analog, chelator, type of radioisotope, and amount of the radiopharmaceutics which is given to the patient.[25] Table 2 summarizes factors resposible for somatostatin analogs accumulation in cancer tumors.

Perhaps the most important factor which is responsible for successful detection and diagnosis of human cancer tumors is the proper procedure used for scintigraphy. Three different acquisition techniques are used: the spot scans, the whole body scan or SPECT. Presently, a system which combines the advantages of SPECT functional imaging with the anatomical resolution of CT is implemented. This combined technique allows more precise localization of the possible sites of disease with its morphological and functional characteristics. SPECT/CT technique has much higher specificity for identifying the tumors when compared with SPECT or CT alone.

Scintigraphy with Radiolabeled Peptides

Radionuclides provide a diagnostic modality in which radiopharmaceuticals are utilized for the localization of them in organs and tissues. In control clinical conditions, the application of radiolabeled somatostatin analogs visualizes thyroid gland, spleen, liver, kidneys, gut, urine bladder, and in some cases the pituitary gland (Fig. 2). The accumulation of radiolabeled somatostatin analogs in the pituitary gland, spleen and thyroid gland is a result of natural existence of somatostatin receptors in these organs. Accumulation in the gut results from the efflux of radioactive bile into the intestine. Accumulation in kidneys results primarily from the reabsorption of radiolabeled agent from urine and only in part because some of the cells of urinary

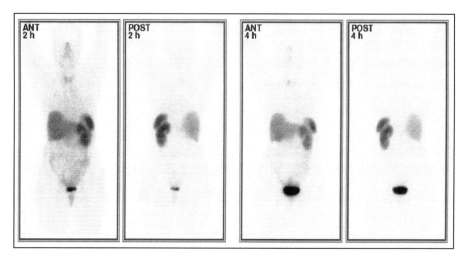

Figure 2. 99mTc-HYNIC-octreotide scans—normal distribution of the radiopharmaceutics.

tract and vasa recta have sst receptors. During the interpretation of scintigraphic results we have to remember that the radiolabeled somatostatin analogs are present in several organs responsible for their elimination: gall bladder, additional spleen, acute brain infarct, in fresh surgical wounds, and in breast and in the chest, when patient is after radiotherapy.[26] As it was already mentioned, the scan can be modified by administration of cold somatostatin analogs, when decreased accumulation of radiopharmaceutics is observed in liver and spleen. This effect is less extended in diseased tissues. The relative increase of uptake in the pathological foci may be a cause of misinterpretation of the result and could suggest the increased density of receptors in the liver metastases.

Diagnostic somatostatin receptor scintigraphy demonstrates the distribution of sst receptors in the whole body. It is therefore possible to localize the primary tumor and metastases, often in unexpected organs or tissues. This examination is able to show the pathologic lesions in size 5-10 mm.

Clinical Indications for Scintigraphy with Radiolabeled Peptides

The fundamental indications for a sst scintigraphy are:

1. Localization of metastases to define staging of cancer process
2. Indication for the treatment at the present stage of cancer process
3. Monitoring of the progress of the therapy
4. Evaluation of the treatment with cold somatostatin analogs
5. Before the treatment with radioisotope
6. Before surgery when chemotherapy will follow it
7. Before the liver transplantation
8. Intraoperational examination for the best surgical outcome
9. In several types of cancer, as a suggestion for the grade of malignancy (high expression of sst receptors is observed in high differentiated tumors, a low expression in malignant tumors).

It is important to note, that scintigraphy (which is made to find sst receptors in vivo) not always gives the same results as in vitro studies and this means that in vitro examinations should not be taken as a final result. For example, in vitro examination shows high expression of somatostatin receptors in normal lymph nods and thymus, but they are visible only sporadically in scintigraphy. In contrast, lymphoma cells have rather small number of sst receptors but they are very often visible in scan.[27] At present, scintigraphy with sst analogs is advisable in tumors like: gastrinoma, glucaconoma, vipoma and carcinoid.

In these tumors sensitivity of other diagnostic tools are very limited. Perhaps, the most successful diagnostic procedures should include several techniques, and the most advisable would be to do SPECT/CT.[28,29]

A Step-Wise Strategy in Patient Management

A high quality of scintigraphy is a fundamental step in proper diagnosis. It is therefore essential to remember about the management and preparation of the patient before the examination. Twenty four hrs before the examination enema is advised. In some patients the examination takes place during the course of therapy with somatostatin analogs and then is important to withdraw the medication 24 hrs before. The next dose could be give even 4-6 hrs after the examination. In the group of patients treated with LAR, the examination should be made directly before the next dose of treatment. Several authors stated that LAR treatment does not change the scintigraphy results and the examination could be done at any time of treatment. The strategy for the examination is always related to particular radiopharmaceutics.

One portion of 111In-Octreoscan for single use contains 10 μg of peptide with 200 MBq radioactivity. This amount of radioactivity is adequate for planar and SPECT examinations. In the examinations with 111In the medium-energy collimator should be used. The energy window should be centered over both 111In photon peaks (172keV and 245keV) with the window width of 20%. Because of a relatively long effective half-life and interfering background radioactivity planar and SPECT studies are preferably performed 24 h after injection of radiopharmaceuticals. Presently, scintigraphy with the 99mTc-tracer have increased meaning and use in practice. Patients, before administration of radiopharmaceutics are advised to drink fluid and to be well hydrated. The preferred radioactivity is 760MBq. The examination takes place 90 min after radiopharmaceutics administration and at first AP and PA scans of whole body should be made. 120-180 min after radiopharmaceutics administration chest SPECT and abdomen should be made. A low-energy collimator should be used. Energy windows should be centered at the 140keV with the window width of 20%. The scans must be interpreted together with CT.

The interpretation of scintigraphy is a complex process and has to answer for the most important question: is the hot spot a typical picture of cancer tumor?. To get the most precise answer, the tumor/background ratio of the radioactivity must be made in two windows of time: 4 and 24 hrs after radionuclide administration. Accumulation of radioactivity is not specific if the tumor/background ratio decreases or stays at the same level (e.g., increased blood flow or increased efflux is a case of hot spot). In some NET tumors transient, unspecific accumulation of sst analogs also can be observed.

Tumors Identification and Surveillance

Pituitary Adenomas

Octreotide scintigraphy of the pituitary visualizes tumors which produce GH, TSH and also tumors without hormonal activities. Pituitary tumors producing ACTH are not visible. On the contrary, ACTH producing tumors located besides the pituitary are characterized by the existence of sst receptors and therefore octreotide scintigraphy is the method of choice for their diagnosis.[30] Following pathological processes located in the region of sella turcica should be taken for differentiation diagnosis: lymphoma, granulomatous diseases, meningioma and metastases of tumors. Octreotide scintigraphy in acromegalic patients supposes to help make a decision about the proper therapy; however, many studies do not confirm this thesis and currently scintigraphy is not advisable in these cases. Similarly, at present there is no indication for diagnosis and therapy of the hormonally non active pituitary tumors.[31,32]

Lymphomas

Lymphomas have a low density of somatostatin receptors and the sensitivity of scintigraphy is limited. In most cases one or several spots of increased accumulation of radiolabeled tracer are visible, but a degree of its accumulation is much lower than in NETs. In cases of low-grade NHL positive results could be reached in 84% subjects. In 20% of patients scintigraphy localizes the pathological changes, even when other techniques did not show anything. In 38% of cases the radioisotopic examination was false positive.[33,34] Therefore scintigraphy is advisable only in particular group of patients and needs clinical verification. The evaluation of lesion related sensitivity shows that 94% spots have sst receptors, from those 98% have them over the diaphragm and 67% under the diaphragm. Therefore scintigraphy is more precise in the diagnosis of all spots located over the diaphragm. In 18% of patients from this group the localization of pathologic changes advanced in I and II degree could lead to the modification of the treatment and is very important for the future of the patients.

Paragangliomas

Almost all of these tumors are characterized by a high expression of sst receptors and they can be visualized by scintigraphy with somatostatin analogs. Furthermore, scintigraphy can show the pathological changes very early (in 10% of patients it shows the multiple tumors or metastases). Kweekboom et al[35] showed in 25 patients the additional pathologic spots in 36% patients.

Thyroid Cancer

For the diagnosis of medullary thyroid carcinomas several radiopharmaceutics are used: somatostatin analogs, MIBG, 99mTc-V-DMSA, 99mTc-MIBI and 201Tl. Sensitivity of scintigraphy with use of octreoscan is 50-70%. Kweekboom et al[36] showed that the ratio of calcitonin to CEA is high in group of patients with positive results of scintigraphy after administration of somatostatin analogs. These authors suggest that expression of sst receptors correlates with high differentiation of cancer. Therefore, scintigraphy can play the prognostic role.[37] Sensitivity of the radionuclide scan is also size dependent, and in small size tumors the results are questionable.[38] In all cases of the thyroid cancer FDG-PET examinations are not advisable, similarly like in other tumors with a slow growth.

Several observations suggest that scintigraphy with sst analogs could improve the diagnostic process in cases of papillary, follicular as well as anaplastic thyroid carcinomas.[39,40] Scintigraphy is also helpful in localization of metastases, even if they do not accumulate 131I. It is not necessary to withdrew thyroxine before examination. The implementation of scintigraphy with sst analogs could improve diagnosis and therapy in this group of patients.

Small Cell Lung Carcinoma

Scintigraphy with 111-In Octreoscan is able to localize almost all primary small cell lung carcinoma (SCLC) tumors. The only false negative results come from metastases with extremely low expression of sst receptors. Scintigraphy could show unknown spots of cancer and metastases to the brain.[41,42] Results published by Kweekboom et al[41] show that scintigrapy is more precise in the evaluation of the status of disease in 36% of patients. The diagnose of even small changes in the brain could lead to much earlier radiotherapy and could increase success of treatment.[43] However, several authors deny the applicability of scintigraphy in this group of patients.[44] Similar meaning has the use of depreotide with 99mTc. It has an affinity to receptors sst2, sst5 and sst3. Sst3 receptors are present in small cells and in non small cells tumors (NSCLC), therefore scintigraphy is able to visualize different types of lung cancers. Sensitivity of scintigraphy is 96% and is comparable to PET. Specificity is 74%, independently of the type of tumor. Similarly to the examination made with Octreoscan, not all metastases of small cell lung cancer have the expression of sst receptors and than they could be invisible to depreotide.[41-50] It is noteworthy, that in NSCLC the mechanism of the visualization of tumor is different: tumor cells do not have sst.[41,45] However, sst receptors are present in activated leukocytes and on proliferating neuroendocrine cells, which are located near the primary cancer.

Breast Cancer

Somatostatin receptor scintigraphy localized 75% of primary breast cancers.[46] Scintigraphic examination could show not only pathological process but also the existence of cancer cells in not enlarged lymph nodes, and additionally it could be used to show the recurrence of cancer process after the treatment.

Gastrointestinal Tumors

Neuroendocrine tumors located in the gut are characterized by a very high expression of sst2 receptors; therefore the radiopharmaceutics of the first choice is Octreoscan. This type of examination is able to localize the primary tumor and also its metastases.[47] Additionally, the results of this examination could be used for the evaluation which strategy of the treatment should be given to the patients. If only a single localization is present, the surgical treatment is advisable, the existence of metastases shows the need for other type of treatment, including somatostatin analogs (Fig. 3).

Carcinoid

The sensitivity of scintigraphy is 80-100%.[48-50] Several authors pointed out that only scintigraphy shows cancer spots which were invisible in the examinations made by other techniques. The results of scintigraphy should be the basis for the decision which treatment to be used and also could show the results of the treatment. It was shown that in the group of patients with spread cancerous process the degree of expression of somatostatin receptors predicts the treatment success. The positive results of scintigraphy should be an indication for the use of somatostatin analogs, and the negative result should lead to the use of chemotherapy (Figs. 4-6).[51]

Pancreatic Tumors

The majority of pancreatic tumors expresses the sst receptors. Scintigraphy should be used when other techniques are not successful in localization of the tumor. The european multicenter studies[52] have shown that the sensitivity of Octreotide scintigraphy: in glucagonoma it was 100%; in gastrinoma, 60-90%; in lymphoma, 88%; and in nonfunctioning islet cell tumors, 82%. If

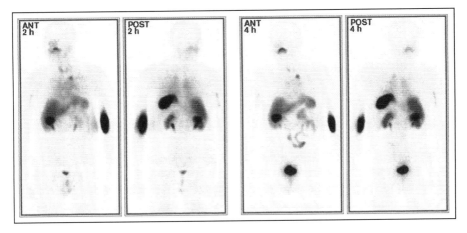

Figure 3. 99mTc-HYNIC-octreotide scans in patient with NET and metastasis in the left lung.

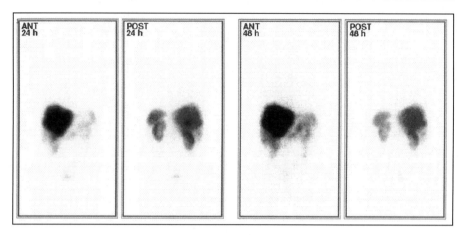

Figure 4. 99mTc-HYNIC-octreotide scans in patient with neuroendocrine tumor of the liver.

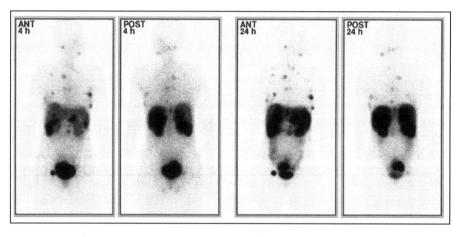

Figure 5. 111In- octreotide scans in a patient with multiple carcinoid metastases. The examination was performed before qualification of the patient to the targeted radionuclide therapy with 90Y labeled somatostatin analog.

the pancreatic tumor is not visualized it does not mean that this tumor have not somatostatin receptors. Scintigraphy has some limitations and its low sensitivity found in some tumors may be related to a size and location of the tumor or to scanning procedure. The sensitivity of other techniques for visualization of NET/GEP tumors in pancreas (like MR, CT, USG, and angiogram) is only 50%.[53] The endoscopic ultrasound is of a great value in these cases.[54,55] Gibril et al[56] showed that sensitivity of scintigraphy in Zollinger-Ellison syndrome is very high. Thus, scintigraphy was a lowest cost examination with good sensitivity in comparison to the other imaging techniques combined. Importantly, scintigraphy should be a major factor for the decision about the therapy. Lebtahi et al[56] showed in the group of 160 patients that scintigraphy results changed the therapy in 25% of cases and in Termanini et al[58] studies with gastrinoma patients this factor was 47%. Scintigraphy results should replace calcium test in gastrinoma and arginine test in glucagonoma.

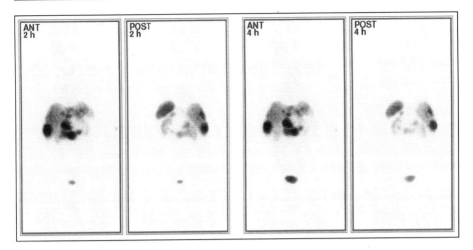

Figure 6. 99mTc-HYNIC-octreotide scans in a patient with jejunum carcinoid and metastases to the liver.

Insulinoma

Seventy % of insulinomas have sst2 and sst5 receptors, 50% of these tumors have sst1, and in 15-20% of insulinomas sst3 and sst4 were found.[59] The coexpression of sst5 and sst2 in majority of insulinomas could lead to heterologous receptor dimerization. The sensitivity of scintigraphy with administration of Octreoscan is approximately 50-60%.[45,60] Because of this very low sensitivity the best method for insulinoma localization is ASVS test with calcium and intraoperative ultrasound.

Neuroblastoma/Pheochromocytoma

These tumors, characterized by a high expression of sst2a receptors, can be detected in approximately 90% of cases. Patients bearing neuroblastomas that express somatostatin receptors have a longer survival than those with receptor-negative tumor. This examination has some limitations because of the high kidney uptake of Octreoscan. Therefore for the diagnosis of neuroblastoma and pheochromocytoma MIBG scintigraphy is preferable.

For the identification of metastases the use of MIBG scintigraphy lead to detection of 90% of lesions with diameter higher than 1 cm, when sst scintigraphy shows only 25%. Thus, both methods should be complementary; the indication for sst scintigraphy is a negative result of 123I MIBG scintigraphy in these patients in whom metastases are highly probable.

Hepatocellular Carcinoma

Recent reports suggest a possible neuroendocrine component in hepatocellular carcinomas (HCC)[61] and that the patients with HCC treated with octreotide survived longer time than patients who had not been treated.[62] Reubi et al[63] demonstrated that somatostatin receptors and VIP receptors are expressed in 41% and 47% of cases of HCC, respectively. No correlation was seen between receptor status, and the type and grade of malignancy of the tumors. Somatostatin receptor density is low in approximately half of the receptor positive cases of HCC. In summary, somatostatin receptor density in HCC is low compared with the much higher receptor density in metastases from neuroendocrine tumors into the liver. There are no data concerning the scintigraphic results in this group of patients.

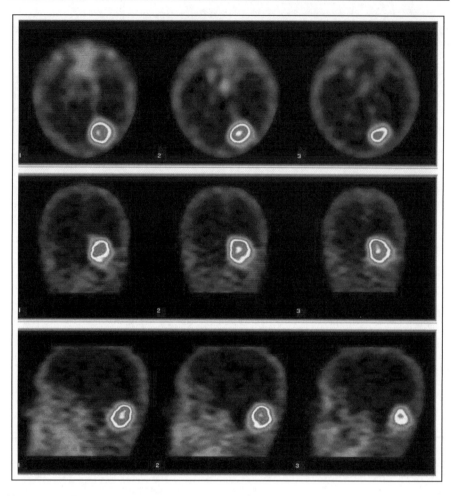

Figure 7. 99mTc-HYNIC-octreotate scans (SPECT technique) in a patient with cerebellar glioma.

Brain Tumors

A high incidence of sst receptors has been reported in some tumors of central nervous system. Sst2 receptor is the most abundantly expressed.[64] Somatostatin receptors have been identified in meningiomas, neuroblastomas, medulloblastomas, and as well as in differentiated astrocytomas, but not in undifferentiated glioblastomas.[65] Majority of these tumors express a high density of sst receptors as compared to surrounding tissues which allows them to be readily visualized in scintigraphy. This fact that some of brain tumors contain sst receptors was an indication for the introduction of the treatment with radiolabeled somatostatin analogs (Figs. 7,8).

Positron Emission Tomography (PET)

Positron emission tomography assesses the function of different metabolic pathways in the tumors. It offers the scans with high sensitivity and good resolution. Therefore recently special emphasis has been given to the development of peptides labeled with positron emitters. Clinical studies were performed with 68Ga-DOTA,Tyr(3)-octreotide.[66] Sensitivity of this method seems to be better than classical sst scintigraphy.[67,68]

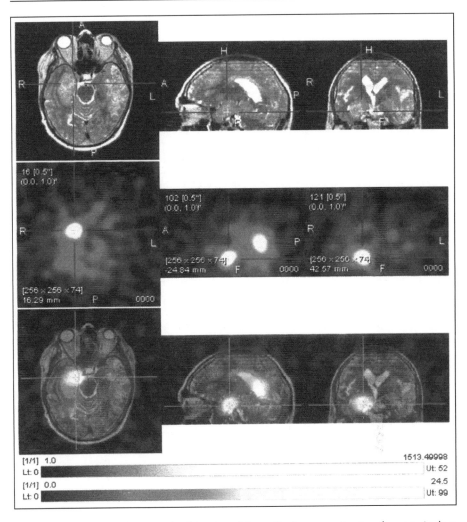

Figure 8. 90Y-DOTATATE scans after injection of the radiopharmaceutics in a therapeutic dose to the brain tumor (glioma).

18F-labelled deoxyglucose (FDG) was the first positron tracer used for evaluation of NET/GEP.[64] FDG is an analog of glucose in the human body and the intensity of its accumulation in the tissue is reflecting the malignancy of the tumor. However, since the majority of GEP/NET tumors are well-differentiated, slow-growing tumors, they have a low metabolic rate and cannot be visualized efficiently with FDG. FDG uptake is significantly increased in poorly-differentiated tumors without somatostatin receptors.[70] The other tracer used in the visualization of GEP is 5-hydroxytyptophan (5-HTP), where the metabolic pathway of synthesis of 5-hydroxytryptamine occurs. 11C-5-HTP is specifically trapped by serotonin-producing tumors and therefore PET imaging with 5-HTP has been shown to be superior compared with CT scanning in diagnosing GEP tumors and monitoring their therapy. PET imaging with FDG or 5-HTP is a complimentary measurement in this group of patients where scintigraphy with sst analogs is negative.

Recently, based on the ability of NET to store the biogenic amines, 18F-DOPA as a positron radiopharmaceutics is evaluated. It seems that FDOPA PET performs better than sst scan in visualizing NETs.[71] In NET tumors located in pancreas first positive results were obtained with l-DOPA and FDG.

In the group of patients with suspected metastases into bones the scintigraphy with MDP is the standard procedure.

Radionuclide Therapy

The discovery of somatostatin and somatostatin receptors created a possibility for therapy of NET. Application of cold somatostatin analogs give positive results in management of several symptoms related to the tumors. Therapy is well tolerated by patients. Importantly, sst analogs besides their influence on malignancy symptoms, have also an important antiproliferative activity, which was shown in experimental studies with various tumor cells, and additionally they influence angiogenesis.[27] These specific effects are not repeated in humans, as yet. In stomach cancer it was shown that the long therapy with somatostatin analogs stabilized the process in 50% patients.[72] However, this effect was much smaller in breast cancer, lung cancer and prostate cancer. The natural next step after the peptide receptor scintigraphy is radiotherapy. The principle for the therapy is that the analog serves as a vehicle for radioisotope emitter with the radiation energy able to destroy the cancer cells. The critical factor for such action is the application of a proper sst analog with a proper radioisotope.

The other option for cancer treatment is the application of sst analogs together with cytostatic substances.[73] Although this kind of medication was not used in humans but experimental results have shown that the most promising is AN-238 (2-pyrrolinodoxorubicin together with octapeptide RC-121.

Radiobiologic Principles in Radionuclide Therapy

Radiotherapy, brachytherapy and targeted radiotherapy belong to the same group of therapeutic modalities, however, they differ substantially.

1. Radiotherapy and brachytherapy is based on photons whereas radiations in radionuclide therapy are based on α/β particules.
2. The dose of radiation in targeted radiotherapy is lower, but the exposition time is longer and in conventional radiotherapy the radiation is high in relatively short time.
3. In radionuclide therapy the uptake of radioisotope is heterogeneous, because the density of the receptors and internalization of radiopharmaceutics are not homogenous.
4. In most cases of radionuclide therapy radiation is absorbed not in the cell where radiopharmaceutics is coupled, but in the surrounding tissues.
5. Linear energy transfer for beta particles (the most used radiation in radioisotopes therapy) is relatively low and amount in about 0.2 keV/Um.

All mentioned factors indicate that the key factor for successful radioisotope therapy is high concentration of radiopharmaceutics in the malignant tumor. The main target for biological effects of radiation is DNA of cancer cells. Radiation can cause several toxic effects in DNA: single-strand break, double-strand break, base damage, DNA-protein cross-links and multiply damaged sites. The majority of these damages could be repaired in a short time period, with exception of double and multiply damaged sites. Almost all these damages are caused by increased activity of free radicals, and to less extend caused by direct disruption of DNA. The degree of destruction is related to the energy of radiation.

Results of Preclinical Data

The therapy with radiolabeled somatostatin analogs was introduced after a long period of experimental work. In mice, it was shown by Zamora et al[74] that injection of 188Re-RC-160 directly into tumor possesing sst receptors leads to 90% of lesion reduction. The intravenous administration of [90Y-DOTA-D-Phe,Tyr3] octreotide in patients with pancreatic tumor leads to a successful tumor's remission in 5 cases from 7.[75] The influence of [111In-DOTA0]-octreotide was observed by Slooter et al[76] in sst positive and negative liver tumors. They have shown that application of radiopharmaceutics caused a significant reduction of metastases but only in sst positive-tumors . Similar effects have been shown in pancreatic tumors in rats (CA 20948) after the injection of [177Lu-DOTA,Tyr3] octreotate.[77] It was shown that tumors with diameter smaller than $1 cm^2$ are destroyed almost in 100%, whereas the tumors with a higher diameter are destroyed only in 40-50%. The efficacy of therapy depends on the density and the type of receptors. Recently, the promising results were published with a mixture of radionuclides 90Y and 177Lu.[78] It was suggested that 90Y radioisotope is more effective in larger tumors, when 177Lu is more effective in smaller tumors. It was also shown that administration of both radionuclides leads to longer survival of experimental animals compared with animals treated only with 90Y or only with 177Lu.[79]

It is well known that the effectiveness of radiotherapy depends on tissue sensitivity to radiation. The tissue sensitivity to radiation is related to the existence of suppressor gene Bax and wt p53. Recent studies in human breast cancer cells demonstrates that both of them are being activated through sst receptors, achieved by octreotide. It is suggested that octreotide tagged radionuclides should elicit cytotoxic response restricted not only to DNA damage but also to the triggering of apoptosis.[80]

Somatostatin Analogs Used in Targeted Radiotherapy

The first somatostatin analog used in the treatment of NETs was [111In-DTPA] octreotide. It was given in several doses, up to the maximal dose of 160 GBq. Presently, the most prevalently used analog is 90Y-DOTA Tyr3-octreotide (90Y-DOTATOC)[81-83] and OCTREOTATE, which has higher affinity to somatostatin receptors.[10] The primary indications for the therapy are multifocal carcinoid and pancreatic tumors (Fig. 9). Some other tumors could be treated also by this method. Until now there is no confirmation for the usefulness of this therapy in lymphomas or nonsmall cell lung carcinoma, although these tumors show a moderate affinity for SST but a high sensitivity for radiation.

Krenning et al[83] used 111In DTPA octreotide in 30 patients in advanced stage of cancer, with the maximal cumulative patient dose 74 GBq. They showed a significant reduction of symptoms, reduction of tumor size and hormones secretion. In the next paper[84] the authors studied 40 patients; they used a higher dose of 111In-DTPA octreotide (20 GBq - 160 GBq). However, they observed a reduction of symptoms only in 8% of patients. It was shown that

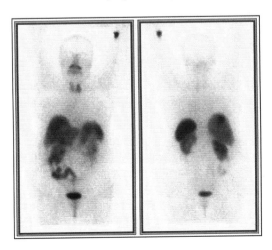

Figure 9. 99mTc-HYNIC-octreotate scans in a patient with a colon carcinoid and metastases to the liver. Examination was performed before qualification of the patient to the targeted radionucline therapy with 90Y-DOTATATE.

therapeutic effects correlated with tumor size and accumulation of tracer in the tumor.[85] Therefore it is possible that results obtained in this group were related to the fact that therapy was performed in the advanced stage of cancer process.

It is important to state, that radionuclide therapy has some side effects. When the dose was 100 GBq, transitory toxic effects from bone marrow were observed: myelodysplastic syndrome and lower count of leukocytes, therefore 100 GBq is considered to be the maximal tolerable dose. It was suggested that the therapeutic effect of [111In DTPA]-octreotide is related to the emission of Auger electrons having a low tissue penetration. Its effect is limited practically only to these cells which internalize the receptor-radioligand complex. The radiation emitted from the receptor positive cells therefore cannot to kill the neighboring receptor negative cells in tumors with receptor heterogeneity. Consequently, new somatostatin analogs that can be labeled with beta emitting radionuclides were developed. This resulted in development of [DOTA,Tyr3] octreotide (DOTATOC), DOTA-Lanreotide (DOTALAN) and [DOTA,Tyr3]Octreotate (DOTATATE) which could be labeled by 90Y or 177Lu. The first radionuclide investigated was 90Y emitting beta-particles with the high maximum energy (2.27 MeV) and a long maximum particle range (10 mm). The first tracer labeled by beta emitter was DOTA-chelated somatostatin analog DOTATOC labeled by 90Y, which was highly hydrophilic and had a higher affinity to sst2 as compared to octreotide.[86] The next tracer was lanreotide labeled with 90Y, which is highly lipophilic and has a higher affinity to sst2 and sst5. Some therapeutic trials have been performed using different somatostatin analogs with beta emitting radionuclides. Otte et al[82] described 29 patients who received injections of 90Y-DOTATOC. Twenty patients showed disease stabilization; two had partial remission; four of them had a reduction of tumor size less than 50%; and only in three of them a progression of tumor growth was observed. The cumulating dose was 6.1 ± 1.34 GBq/m^2. Paganelli et al[87] treated 30 patients with 90Y-DOTATOC. Full or partial remission was observed in 23% patients, 64% patients showed disease stabilization, in 13% of them a progression of tumor growth was noted. Valkema et al[88] described 22 patients with neuroendocrine tumors observed during 14 months. 10 patients had disease stabilization, 10 of them had improvement in clinical symptoms and in 12 of them was no change in their symptoms. Waldheer et al (Basel group)[81,89] reported two studies in patients with NET. In the first study[89] 39 patients with GEP and bronchial tumors treated with 90Y-DOTATOC (7.4 GBq/m^2 injected at 6 weeks intervals). They showed the objective response rate according to WHO criteria in 23% of patients. In pancreatic tumors (13 patients) the response rate was 38%. A complete remission were found in 5 %, a partial remission in 18%, a stable disease in 69%, and a progressive disease in 8%. A significant reduction of clinical symptoms was observed in 83% of patients with diarrhea, in 46% of patients with flush, in 63% of patients with wheezing, and in 75% of patients with pellagra. The overall clinical benefit was observed in 63% of cases. Side effects (grade 3 or 4) were observed in some patients: lymphocytopenia in 23%, grade 3 anemia in 3%, and grade 2 renal insufficiency in 3% of cases. In the next study[86] two protocols were compared: in one, patients received four injections of 1,850 MBq/m^2 at intervals of 6 weeks and in second, two injections of 3,700 MBq/m^2 were administered at an interval of 8 weeks. An increased percentage of complete responses and partial remission was noted in the second group of patients (24% versus 34%), while side-effects were not significantly different. In conclusion authors indicated that treatment interval and dosing may play an important role in the outcome of therapy. Chinol et al (Milan group)[90] treated a total of 256 patients with 90Y-DOTATOC by using two distinct protocols with and without the administration of kidney protecting agents (amino acids infusion). Eighty percent of the patients presented the progressive disease before the start of therapy.

The patients received three or more equal i.v. injections of 90Y-DOTATOC, starting with 1.1 GBq per cycle in escalating dosage in subsequent groups. Administration consisted of 130 µg of DOTATOC labelled with 4.81 GBq of 90Y. The cumulative activity ranged between 7.4 and 21.3 GBq. In 21% of patients a partial or complete response and in 44% of patients a stability of disease were documented. Forty eight percent of patients presented nausea or vomiting after amino acids infusion. No major acute reactions occurred with 90Y-DOTATOC application up to the activity of 5.55 GBq per cycle. Reversible grade 3 hematologic toxicity on white blood cells and/or platelets was found in 3 of 7 patients injected with 5.18 GBq, which was defined as the maximum tolerated dose per cycle. Lymphocyte toxicity (grade 3 or 4) was observed in 77.5% of patients. There were not acute or delayed nephropathy. Aim of the another studies (Rotterdam group)[91,92] was to define the maximal tolerated single- and four-cycle doses of 90Y-DOTATOC (Rotterdam group). Fourty seven patients were evaluated, in all of them amino acids administration was added as a kidney protective agent. Before the treatment 81% of the patients had the progressive disease. With cycle doses ranging from 1.3 to 10.8 GBq and cumulative doses from 1.7 to 27 GBq, the maximal tolerated dose was not reached. Three patients had dose-limiting toxicity. Minor remission were seen in 18% of patients, and a partial response in 10%. A clear dose-response relation was documented in this study: the percentage reduction of tumor volume increased with the increasing tumor radiation dose up to 600 Gy.[93]

90Y-DOTALAN treatment was evaluated in the MAURITIUS trial.[94] The cumulative treatment dose of up to 8.58 GBq 90Y-DOTALAN were given as a short-term intravenous infusion. In summary 154 patients were treated; minor response was observed in 14% of patients, and stabilization of the disease in 41% of patients. Potential indications for 90Y-DOTALAN treatment in radioiodine-negative thyroid cancer, hepatocellular cancer, and lung cancer were suggested. There were no severe or chronic hematological toxicity or renal/liver function parameters changes due to treatment. DOTATOC showed a higher tumor uptake than DOTALAN in most patients.

Recently new somatostatin analog [Tyr3]octreotate is included to the targeted radiotherapy. Octreotate has a very high affinity for sst2 and a high tumor uptake was documented in sst2-positive tumors in clinical studies. Additionally a very good of tumor-to-kidney uptake ratio was noted. This analog has a ninefold higher affinity for the somatostatin receptor subtype 2 as compared with octreotide.[10] The use of 177Lu for peptide labeling was shown to be very successful in promoting tumor regression and lengthening survival in a rat model.[73] 177Lu emits gamma radiation with suitable energy for imaging and beta-particles with low to medium energy (maximum 0.5 MeV). The range of the beta-particles is 2 mm. Accordingly, less cross-fire induced radiation damage in the renal glomeruli can be expected and higher percentage of the 177Lu radiation energy will be absorbed in small tumors and micrometastases (Fig. 10). The effects of 177Lu-Octreotate therapy have been reported recently in 18 patients with a variety of progressive NETs, with 39% of patients showing a partial remission with minimal adverse effects.[88] In the next study[95] the effect of 177Lu-DOTATATE were studied in 35

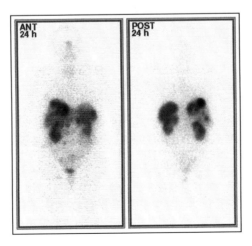

Figure 10. 177Lu - DOTATATE post-therapeutic scans in this same patient (90Y/177Lu-DOTATATE was applied in a dose of 3.7 GBq).

patients with GEP. Patients were treated with dosages of 3,700, 5,500 or 7,400 MBq up to the final cumulative dose of 22.2-29.6 GBq with intervals of 6-9 weeks. Complete or partial response were found in 38% of the patients. The side-effects of treatment were few and transient, with mild bone marrow depression. Kidney function did not deteriorate in any patient. Because of the limited efficacy of alternative therapies the authors concluded that radionuclide therapy should be useful in patients with GEP without waiting for tumor progression. Similar results were obtained in the group of 131 patients.[96] only one patient developed the renal insufficiency, and another patient developed the hepatorenal syndrome. Hematologic toxicity (grade 3 or 4) was observed in in less than 2% of patients. Complete remission was observed in 3 patients, partial remission in 32 patients, minor response in 24 patients and stable disease in 44 patients. Higher remission rates were positively correlated with high uptake on pretherapy somatostatin receptor imaging and limited number of liver metastases.

Successful therapy with somatostatin radiolabeled analogs is dependent from several factors:

1. Size of tumor
2. Morphology of tumor
3. Distribution, density and heterogeneity of sst receptors
4. Angiogenic vessels and peritumoral vessels
5. Tumor tissue radiosensitivity
6. Various degree of internalization of the receptor-radioligand complex

Toxicity of Radionuclide Therapy

The application of a radioisotope to tumor cells could deliver an effective radiation dose to the tumor; however, it could have some side effects also. It is mainly related into accumulation of somatostatin analogs in normal, healthy organs: bone marrow, kidneys and liver. These organs have somatostatin receptors and they are the sites of elimination for free peptides. Therefore, it is important to restrict the radiation to localized tumor and to keep checking the radiation and to follow up a strict protocol for the administration of radionuclide.

In case of DOTATOC it is kidney toxicity. More than 50% of this tracer is eliminated in 4 hrs after its application and more than 70% is removed by kidneys in first 24 hrs. No degradation product was found in circulating blood. The cumulative renal absorber dose must be limited till 27Gy.[84] It was shown that the infusion of amino acid solution (mostly lysine and arginine) could significantly decrease tracer accumulation in kidneys even by 30-50%[97] and could lower the toxicity. Amino acid infusion does not change the accumulation of sst analogs in tumors. It is important to remember that amino acid injection could cause nausea, vomiting, metabolic acidosis and hyperkaliemia. There was not a noticeable change in the function of pituitary gland. A significant decrease of testosterone levels with increased levels of LH and FSH was shown at the some time. The transient changes in spermatogenesis could be also shown..

Future Perspective for the Therapy

Long-term survival of patients is the primary goal when considering therapy with radionuclides. The new developments for the potential usefulness in this therapy should consider the following factors:

1. Peptide receptors—it was suggested that other peptides could be used in future if there are peculiar receptors in the examined tumor. For example, in breast cancer and prostate cancer we could use two peptides: bombesine and NPY-1. In tumors where sst2 and sst3 exist we could use 1NaI 3 octreotide. This peptide could be used if octerotide and octreotate show a small accumulation in the tumor.
2. Application of various radioisotopes—it was shown in preclinical studies that therapeutic effect in rats was much better when 177Lu and 90Y were used. These studies also show that 90Y is more effective in large size tumors than 177Lu, which gives the better results in small tumors.

References

1. Reubi JC, Landolt AM. High density of somatostatin receptors in pituitary tumors from acromegalic patients. J Clin Endocrinol Metab 1984; 59:1148-51.
2. Shapiro B, Sisson JC, Schulkin BL. The current status of meta-jodobenzylguanidyne and related agents for the diagnosis of neuroendocrine tumors. Quartely J Nucl Med 1995; 39:3-8.
3. Troncone L, Rufini V. 131MIBI therapy of neural crest tumours. Anticancer Research 1997; 17:1823-31.
4. Jenkins SA, Kynaston HG, Davies ND et al. Somatostatin analogs in oncology: A look to the future. Chemotherapy 2001; 47:162-96.
5. Boerman OC, Oyen WJG, Corstens FHM. Radio-labeled receptor-binding peptides: A new class of radiopharmaceuticals. Semin Nucl Med 2000; 30:195-208.
6. Kwekkeboom D, Krenning EP, de Jong M. Peptide receptor imaging and therapy. J Nucl Med 2000; 41:1704-13.
7. Jensen RT, Gibril F, Termanini B. Definition of the role of somatostatin receptor scintigraphy in gastrointestinal neuroendocrine tumor localization. Yale J Biol Med 1997; 70:481-500.
8. Gibril F, Reynolds JC, Chen CC et al. Specificity of somatostatin receptor scintigraphy: A prospective study and effects of false-positive localization on management in patients with gastrinoma. J Nucl Med 1999; 40:539-53.
9. Gabriel M, Muehllechner P, Decristoforo C et al. 99mTc-EDDA/HYNIC-Tyr(3)-octreotide for staging and follow-up of patients with neuroendocrine gastro-entero-pancreatic tumors. QJ Nucl Med Mol Imaging 2005; 49:237-44.
10. Reubi JC, Schaer JC, Waser B et al Affinity profiles for human somatostatin receptor sst1-sst5 of somatostatin radiotracers selected for scintigraphic and radiotherapeutic use. Eur J Nucl Med 2000; 27:273-82.
11. Menda Y, Kahn D. Somatostatin receptor imaging of nonsmall cell lung cancer with 99mTc depreotide. Sem Nucl Med 2002; 32:92-96.
12. Virgolini I, Leimer M, Handmaker H et al. Somatostatin receptor subtype specificity and in vivo binding of a novel tumor tracer, 99mTc NP829. Cancer Res 1998; 58:1850-59.
13. Kwekkeboom DJ, Bakker WH, Kooij et al. [177Lu-DOTA0Tyr3]octreotate: Comparison with [111In-DTPA0]octreotide in patients. Eur J Nucl Med 2001; 28:1319-25.
14. Krenning EP, Bakker WH, Breeman WAP et al. Localization of endocrine-related tumors with radioiodinated analogue of somatostatin. Lancet 1989; 1:242.
15. Lamberts SWJ, Bakker WH, Reubi JC et al. Somatostatin receptor imaging in the localization of endocrine tumours. N Engl J Med 1990; 323:1246.
16. Breeman WA, Kwekkeboom DJ, Kooij PP et al. Effect of dose and specific activity on tissue distribution of indium-111-pentetreotide in rats. J Nucl Med 1995; 36:623-27.
17. Hukowic N, Panatta R, Kumar U et al. Agonist-dependent regulation of cloned human somatostatin receptor types 1-5 (hSSTRI-5): Subtype selective internalization or upregulation. Endocrinology 1996; 137:4046-49.
18. Rocheville M, Lange DC, Kumar U et al. Subtypes of the somatostatin receptor assamble as functional homo- and heterodimers. J Biol Chem 2000; 275:7862-69.
19. Janson ET, Westlin JE, Ohrwall U et al. Nuclear localization of 111In after intravenous injection of [111In-DTPA-D-Phe 1]-octreotide in patients with neuroendocrine tumors. J Nucl Med. 2000; 41:1514-1518.
20. Hipkin RW, Friedman J, Clark RB et al. Agonist-induced desensitization, internalization, and phosphorylation of the sst2A somatostatin receptor. J Biol Chem 1997; 272:13869-76.
21. De Jong M, Bernard BF, Bruin E et al. Internalization of radiolabelled [DTPA0] octreotide and [DOTA0,Tyr3] octreotide:peptides for somatostatin receptor-targeted scintigraphy and radionuclide therapy. Nucl Med Commun 1998; 19:283-8.
22. Koening JA, Kaur R, Dodgeon I et al. Fates of endocytosed somatostatin sst2 receptors and associated agonists. Biochem J 1998; 336:291-298.
23. Ronga G, Salerno G, Procaccini E et al. 111In-octreotide scintigraphy in metastatic medullary thyroid carcinoma before and after octreotide therapy: In vivo evidence of the possible down-regulation of somatostatin receptors. Q J Nucl Med 1995; 39:134-136.
24. Dorr U, Wurm K, Horing E et al. Diagnostic reliability of somatostatin receptor scintigraphy during continous treatment with different somatostatin analogs. Horm Metab Res 1993; 27(suppl):36-43.
25. De Jong M, Breeman WA, Bernard BF et al. Tumor uptake of the radiolabelled somatostatin analogue [DOTA, Tyr3]octreotide is dependent on the peptide amount. Eur J Nucl Med 1999; 26:693-8.

26. Kwekkeboom DJ, Krenning EP. Somatostatin receptor imaging. Sem Nucl Med 2002; 32:84-91.
27. Reubi JC. Peptide receptors as molecular targets for cancer diagnosis and therapy. Endocrine Reviews 2003; 24:389-427.
28. Keidar Z, Israel Y. SPECT/CT in tumor imaging: Technical aspects and clinical applications. Sem Nucl Med 2003; 33:205-218.
29. Krausz Y, Keidar Z, Kogan I et al. SPECT/CT hybrid imaging with 111In-pentetreotide in assessment of neuroendocrine tumours. Clin Endocrinol 2003; 59:565-73.
30. de Herder WW, Krenning EP, Malchoff CD et al. Somatostatin receptor scintigraphy: Its value in tumor localization in patients with Cushing's syndrome caused by ectopic corticotrophin or corticotrophin-releasing hormone secretion. Am J Med 1994; 96:305-12.
31. Borson-Chazot F, Houzard C, Ajzenberg C et al. Somatostatin receptor imaging in somatotroph and nonfunctioning pituitary adenomas:correlation with hormonal and visual responses to octreotide. Clin Endocrinol 1997; 47:589-98.
32. Duet M, Ajzenberg C, Benelhadj S et al. Somatostatin receptor scintigraphy in pituitary adenomas: A somatostatin receptor density index can predict hormonal and tumoral efficacy of octreotide in vivo. J Nucl Med 1999; 40:1252-56.
33. Leners N, Jamar F, Fiasse R et al. Indium-111-pentetreotide uptake in endocrine tumors and lymphoma. J Nucl Med 1996; 37:916-22.
34. Lugtenburg PJ, Lowenberg B, Valkema R et al. Somatostatin receptor scintigraphy useful in stage I-II Hodgkin's lymphomas. J Nucl Med 2001; 42:222-9.
35. Kwekkeboom DJ, Van Urk H, Pauw KH et al. Octreotide scintigraphy for the detection of paragangliomas. J Nucl Med 1993; 34:873-8.
36. Kwekkeboom DJ, Reubi JC, Lamberts SWJ et al. In vivo somatostatin receptor imaging in medullary thyroid carcinoma. J Clin Endocrinol Metab 1993; 76:1413-17.
37. Behr TM, Gratz S, Markus PM et al. Anticarcinoembryonic antigen antibodies versus somatostatin analogs in the detection of metastatic medullary thyroid carcinoma: Are caricnoembryonic antigen and somatostatin receptor expression prognostic factors? Cancer 1997; 80:2436-57.
38. Tisell LE, Ahlman H, Wangberg B et al. Somatostatin receptor scintigraphy in medullary thyroid carcinoma. Br J Surg 1997; 84:543-7.
39. Postema PTE, De Herder WW, Reubi JC et al. Somatostatin receptor scintigraphy in nonmedullary thyroid cancer. Digestion 1996; 1(suppl):36-7.
40. Gulec SA, Serafini AN, Sridhar KS et al. Somatostatin receptor expression in Hurtle cell cancer of the thyroid. J Nucl Med 1998; 39:243-5.
41. Kwekkeboom DJ, Kho GS, Lamberts SW et al. The value of octreotide scintigraphy in patients with lung cancer. Eur J Nucl Med 1994; 21:1106-13.
42. Bombardieri E, Crippa F, Cataldo I et al. Somatostatin receptor imaging of small cell lung cancer (SCLC) by means of 111In-DTPA octreotide scintigraphy. Eur J Cancer 1995; 31A:184-8.
43. Bohuslavizki KH, Brenner W, Gunther M et al. Somatostatin receptor scintigraphy in the staging of small cell lung cancer. Nucl Commun 1996; 17:191-6.
44. Kirsch CM, von Pawel A, Grau I et al. Indium 111pentetreotide in the diagnostic work-up of patients with bronchogenic carcinoma. Eur J Nucl Med 1994; 21:1318-25.
45. Krening EP, Kwekkeboom DJ, Bakker WH et al. Somatostatin receptor scintigraphy with [111In-DTPA-D-Phe1]- and [123I-Tyr3]-octreotide: The Rotterdam experience with more than 1000 patients. Eur J Nucl Med 1993; 20:716-31.
46. Van Eijck CH, Krenning EP, Bootsma A et al. Somatostatin-receptor scintigraphy in primary breast cancer. Lancet 1994; 343:640-3.
47. Oberg K, Kvols L, Caplin M et al. Consensus report on the use of somatostatin analogs for the management of neuroendocrine tumors of the gastroenteropancreatic system. Annals of Oncology 2004; 15:966-73.
48. Kwekkeboom DJ, Krenning EP, Bakker WH et al. Somatostatin analogue scintigraphy in carcinoid tumors. Eur J Nucl Med 1993; 20:283-92.
49. Kalkner KM, Janson ET, Nilsson S et al. Somatostatin receptor scintigraphy in patients with carcinoid tumors: Comparison between radioligand uptake and tumor markers. Cancer Res 1995; 55:5801-4.
50. Westlin J, Janson ET, Arnberg H et al. Somatostatin receptor scintigraphy of carcinoid tumors using the [111In-DTPA-D-Phe1]-octreotide. Acta Oncol 1993; 32:783-6.
51. Kvols LK. Medical oncology considerations in patients with metastatic neuroendocrine carcinomas. Semin Oncol 1994; 21(suppl 13):56-60.
52. Krenning EP, Kwekkeboom DJ, Pauwels EK et al. Somatostatin receptor scintigraphy. Nuclear Medicine Annual. New York: Raven Press, 1995:1-50.
53. Lunderquist A. Radiologic diagnosis of neuroednocrine tumours. Acta Oncol 1989; 28:371-2.

54. De Kerviler E, Cadiot G, Lebtahi R et al. Somatostatin receptor scintigraphy in forty-eight patients with the Zollinger-Ellison syndrome. Eur J Nucl Med 1994; 21:1191-97.
55. Zimmer T, Stolzer U, Bader M et al. Endoscopic ultrasonography and somatostatin receptor scintigraphy in the preoperative localization of insulinomas and gastrinomas. Gut 1996; 39:562-68.
56. Gibril F, Reynolds JC, Doppman JL et al. Somatostatin receptor scintigraphy: Its sensitivity compared with that of other imaging methods in detecting primary and metastatic gastrinomas. A prospective study. Ann Intern Med 1996; 125:26-34.
57. Lebtahi R, Cadiot G, Sarda L et al. Clinical impact of somatostatin receptor scintigraphy in the management of patients with neuroendocrine gastroenteropancreatic tumours. J Nucl Med 1997; 38:853-8.
58. Termani B, Gibril F, Reynolds JC et al. Value of somatostatin receptor scintigraphy: A prospective study in gastrinoma of its effect on clinical management. Gastroenterology 1997; 112:335-47.
59. Bertherat J, Tenenbaum F, Perlemoine K et al. Somatostatin receptors 2 and 5 are the major somatostatin receptors in insulinomas: An in vivo and in vitro study. J Clin Endocrinol Metab 2003; 88:5353-60.
60. Proye C, Malvaux P, Pattou F et al. Noninvasive imaging of insulinomas and gastrinomas with endoscopic ultrasonography adnd somatostatin receptor scintigraphy. Surgery 1998; 124:1134-43.
61. Zhao M, Laissue JA, Zimmerman A. "Neuroendocrine"differentiation in hepatocellular carcinomas (HCCs): Immunohistochemical reactivity is related to distinct tumor cell types, but not to tumor grade. Histol Histopathol 1993; 8:617-26.
62. Kouroumalis E, Skordilis P, Thermos K et al. Treatment of hepatocellular carcinoma with octreotide: A randomised controlled study. Gut 1998; 42:442-7.
63. Reubi JC, Zimmermann A, Jonas S et al. Regulatory peptide receptors in human hepatocellular carcinomas. Gut 1999; 45:766-74.
64. Dutour A, Kumar U, Panetta R et al. Expression of somatostatin receptor subtypes in human brain tumors. Int J Cancer 1998; 76:620-7.
65. Reubi JC. Neuropeptide receptor in health and disease: The molecular basis for in vitvo imaging. J Nucl Med 1995; 36:1825-35.
66. Maecke HR, Hofmann M, Haberkorn U. (68)Ga-labeled peptides in tumor imaging. J Nucl Med 2005; 46(suppl 1):172S-178S.
67. Breeman WA, de Jong M, de Blois E et al. Radiolabelling DOTA-peptides with 68Ga. Europ J Nucl Med Mol Imaging 2005; 32:478-85.
68. Kowalski J, Henze M, Schuhmacher J et al. Evaluation of positron emission tomography imaging using [68Ga]-DOTA-D Phe(1)-Tyr(3)-Octreotide in comparison to [111In]-DTPAOC SPECT. First results in patients with neuroendocrine tumors. Mol Imaging Biol 2003; 5:42-8.
69. Eriksson B, Orlefors H, Sundin A et al. Positron emission tomography in neuroendocrine tumours. Italian J Gastroenterology Hepatology 1999; 31(suppl 2):S167-S171.
70. Erikkson B, Bergstrom M, Sundin A et al. The role of PET in localization of neuroendocrine and adrenocortical tumors. Annals of the New York Academy of Science 2002; 970:159-69.
71. Becherer A, Szabo M, Karanikas G et al. Imaging of advanced neuroendocrine tumors with (18)F-FDOPA PET. J Nucl Med 2004; 45:1161-7.
72. Shojamanesh H, Gibril F, Louie A et al. Prospective study of the antitumoral efficacy of long-term octreotide treatment in patients with progressive metastatic gastrinoma. Cancer 2002; 94:331-43.
73. Nagy A, Schally AV. Targeted cytotoxic somatostatin analogs: A modern approach to the therapy of various cancers. Drugs Future 2001; 26:261-70.
74. Zamora PO, Gulhke S, Bender H et al. Experimental radiotherapy of receptor-positive human prostate adenocarcinoma with 188-ReRC-160, a directly-radiolabeled somatostatin analogue. Int J Cancer 1996; 65:214-20.
75. Stolz B, Smith-Jones P, Albert R et al. Somatostatin analogues for somatostatin-receptor-mediated radiotherapy of cancer. Digestion 1996; 57:17-21.
76. Slooter GD, Breeman WA, Marquet RL et al. Anti-proliferative effect of radiolabeled octreotide in a metastases model in rat liver. Int J Cancer 1999; 81:767-71.
77. De Jong M, Breeman WA, Bernard BF et al. [177Lu-DOTA(0),Tyr3] octreotate for somatostatin receptor-targeted radionuclide therapy. Int J cancer 2001; 92:628-633.
78. Krenning EP, de Jong M, Jamar F et al. Somatostatin rceptor-targeted radiotherapy of tumors: Preclinical and clinical findings. In: Lamberts SWJ, Dogliotti I, eds. The Expanding Role of Octreotide: Advances in Oncology. Bristol: Bioscientifica Ltd., 2002:211-23.
79. De Jong M, Krenning E. New advances in peptide receptor radionuclide therapy. J Nucl Med 2002; 43:617-20.
80. Sharma K, Srikant CB. Induction of wild-type p53, Bax, and acidic endonuclease during somatostatin signaled apoptosis in MCF-7 human breast cancer cells. Int J Cancer 1998; 76:259-66.

81. Waldherr C, Pless M, Maecke HR et al. The clinical value of [90Y-DOTA]-D-Phe1-Tyr3-octreotide (90Y-DOTATOC) in the treatment of neuroendocrine tumors: A clinical phase II study. Ann Oncol 2001; 12:941-5.

82. Otte A, Herrmann R, Heppeler A et al. Yttrium-90 DOTATOC: First clinical results. Europ J Nucl Med 1999; 26:1439-47.

83. Krenning EP, de Jong M, Kooij PP et al. Radiolabelled somatostatin analogues for peptide receptor scintigraphy and radionuclide therapy. Ann Oncol 1999; 10:S23-S29.

84. Valkema R, de Jong M, Bakker WH et al. Phase I study of peptide receptor radionuclide therapy with [In-DTPA] octreotide: The Rotterdam experience. Semin Nucl Med 2002; 32:110-22.

85. McCarthy KE, Woltering EA, Anthony LB. In situ radiotherapy with 111In-pentetreotide. State of the art and perspectives. QJ Nucl Med 2000; 44:88-95.

86. De Jong M, Breeman WA, Bakker WH et al. Comparison of (111)In-labeled somatostatin analogues for tumor scintigraphy and radionuclide therapy. Cancer Res 1998; 58:437-41.

87. Paganelli G, Zoboli S, Cremonesi M et al. Receptor-mediated radiotherapy with 90Y-DOTA-D-Phe1-Tyr3-octreotide. Europ J Nucl Med 2001; 28:426-34.

88. Valkema R, Jamar F, Jonard P et al. Targeted radiotherapy with 90Y-SMT487 (Octreo Ther):a phase 1 study. J Nucl Med 2000; 41(suppl):111P.

89. Waldherr C, Pless M, Maecke HR et al. Tumor response and clinical benefit in neuroendocrine tumors after 7.4 GBq 90Y-DOTATOC. J Nucl Med 2002; 43:610-6.

90. Chinol M, Bodei L, Cremonesi M, Paganelli G. Receptor-mediated radiotherapy with 90Y-DOTA-DPhe1-Tyr3-octreotide:the experience of the European institute of oncology group. Sem Nucl Med 2002; 32:141-7.

91. De Jong M, Valkema R, Jamar F et al. Somatostatin receptor-targeted radionuclide therapy of tumors: Preclinical and clinical findings. Semin Nucl Med 2002; 32:133-140.

92. Smith MC, Liu J, Chen T et al. OctreoTher: Ongoing early clinical development of a somatostatin- receptor-targeted radionuclide antineoplastic therapy. Digestion 2000; 62(suppl 1):69-72.

93. Jonard P, Jamar F, Walrand S et al. Tumor dosimetry based on PET 86Y-DOTA-Tyr3-octreotide (SMT 487) and CT-scan predics tumor response to 90Y-SMT487 (OctreoTher). J Nucl Med 2000; 41:111P.

94. Virgollini I, Britton K, Buscombe J et al. In- and Y-DOTA-lanreotide:results and implications of the MAURITIUS tral. Semin Nucl Med 2002; 32:148-55.

95. Kwekkboom DJ, Bakker WH, Kam BL et al. Treatment of patients with gastro-entero-pancreatic (GEP) tumours with the novel radiolabelled somatostatin analogue [177Lu-DOTA(0),Tyr3]octreotate. Eur J Nucl Med Mol Imaging 2003; 30:417-22.

96. Kwekkboom DJ, Teunissen JJ, Bakker WH et al. Radiolabelled somatostatin analog [177Lu-DOTA0,Tyr3]octreotate in patients with endocrine gastroenteropancreatic tumours. J Clin Oncol 2005; 20:2754-62.

97. De Jong M, Rolleman EJ, Bernard BF et al. Reduction of the renal uptake of indium-111-DTPA-octreotide in vivo. J Nucl Med 1996; 37:1388-92.

CHAPTER 8

Somatostatin Analogs—New Perspectives

Gabriela Melen-Mucha* and Marek Pawlikowski

Abstract

Acromegaly and so-called neuroendocrine tumors (carcinoids, pancreatic endocrine tumors) are at present the only routine indications for somatostatin (SST) analogs therapy. However, the progress in the studies on somatostatin receptors (sst) and the development of the numerous new SST analog molecules, creates new possibilities of the therapeutic applications. These possibilities include: (1) the treatment of pituitary tumors other than those connected with acromegaly; (2) the treatment of non endocrine cancers, including so-called receptor-targeted chemotherapy; (3) the treatment of autoimmune/chronic inflammatory disorders, including Graves' ophtalmopathy and rheumatoid arthritis; and (4) the treatment of diabetic retinopathy and senile macular degeneration.

Introduction

Until quite lately the routine application of somatostatin (SST) analogs in therapy included only acromegaly and neuroendocrine tumors of the gut (for details see chapters VI and VII). Nowadays, the preclinical and clinical studies open several perspectives of new therapeutic applications of SST analogs. These perspectives are created both by the discovery of new SST analog molecules and by the proposals of new applications of the "old", classical SST analogs like octreotide and lanreotide, used for many years in clinical medicine. The new molecules include the analogs with increased selectivity to the chosen subtypes of sst1-5 receptors or more "universal" agonists interacting with almost all sst receptor subtypes, like SOM230 which acts on all sst receptor subtypes except sst4.[1] Another category of the new compounds represent chimeric (hybrid) molecules. Their synthesis is based on the observation that sst receptors can heterodimerize and this heterodimerization can result in the altered biological response.[2] The example of such a molecule is BIM-23A387, which simultaneously interacts with sst2 and dopamine D2 receptors.[3]

SST Analogs in the Treatment of Nongrowth Hormone-Secreting and Nonfunctioning Pituitary Tumors

As it has been stated above, only the growth hormone (GH)-secreting pituitary adenomas, manifested by acromegaly, are routinely treated with SST analogs. However, other types of pituitary adenomas also express sst receptors and because of that they are candidates for such a therapy. The trials of the therapy of pituitary adenomas which do not secrete GH with commonly used SST analogs octreotide or lanreotide were not fully successful.[4,5] The negative results depended probably on the fact that the above-mentioned "classical" SST analogs interact mainly with sst2 receptors, whereas PRL and ACTH inhibition by SST is mediated mostly

*Corresponding Author: Gabriela Melen-Mucha—Department of Immunoendocrinology, Medical University of Lodz, Sterlinga 3, 91-425 Lodz, Poland. Email: g.m-mucha@wp.pl

Somatostatin Analogs in Diagnostics and Therapy, edited by Marek Pawlikowski.
©2007 Landes Bioscience.

by subtype sst5.[6,7] However, the new analogs, interacting with other receptor subtypes, are under the investigation. The in vitro effects of an "universal" somatostatin analog SOM230 on PRL secretion from human prolactinomas are encouraging.[1] Another new molecule under the consideration for prolactinoma treatment is a somatostatin/dopamine chimera acting simultaneously on somatostatin receptors sst2 and dopamine receptors D2. The inhibitory effect of this molecule on PRL secretion from mixed GH/PRL secreting adenomas has been recently found.[3] SOM230, which binds (among others) with high affinity to sst5 receptors, is much more effective than octreotide in inhibiting the ACTH release from corticotroph adenomas in vitro.[7] The nonfunctioning pituitary adenomas express mostly sst1, sst2B and sst5 receptors.[8] The recent in vitro studies show that a selective agonist of sst1 exert an inhibitory effect on viability of nonfunctioning pituitary adenoma cells which express this receptor subtype.[9,10] Thus, the SST analogs acting preferentially on sst1 receptors might be useful for medical treatment of this type of pituitary adenomas. The case of sst5 agonists is more controversial because the results of the studies are not concordant; promoting[9] as well as reducing effects[10] on cell viability were found. The considerable inhibitory effect of adenoma cells viability is also exerted by a somatostatin/dopamine chimera.[10] Thus the latter compound might also be a candidate for the medical treatment of nonfunctioning pituitary adenomas.

SST Analogs in the Treatment of Nonendocrine Cancers: Possible Application of New Analogs and Receptor-Targeted Chemotherapy

In addition to paninhibitory effect of SST analogs on the hormone secretion, these substances influence also several processes involved in the growth of normal and neoplastic tissues such as proliferation, apoptosis, angiogenesis, immune system activity, secretion and action of various growth factors and cytokines.

The first report showing antiproliferative activity of SST was published in 1978 by the editor of this book and revealed that SST inhibited the proliferation of normal pituitary cells (the mitogenic action of TRH on the anterior pituitary gland in vitro).[11] A few years later, Mascardo and Sherline,[12] showed that SST inhibited the centrosomal separation and cell proliferation of some tumor cells (HeLa and gerbil fibroma cells). As a consequence of these two findings an enormous number of data have been published concerning anti-growth effect of SST and its analogs on normal and tumoral tissues. At the end of 1980s, the list of neoplasms whose growth processes were changed via SST analogs was very long and included besides various endocrine tumors, which is the subject of another chapter in this book, also several nonendocrine tumors such as pancreatic cancer, breast cancer, prostatic cancer, colon cancer, hepatocellular cancer, renal cancer, laryngeal cancer, meningioma, lymphoma, thymoma, melanoma, small cell lung cancer, osteo- and chondrosarcoma.[13,14] The antitumor effect of SST analogs was observed in both in vivo and in vitro studies (cancer cell lines), including animal and human cancers, spontaneous and carcinogen-induced cancers, cancers transplantable into nude mice and into normal animals.

Moreover, it was found out at the same time that various tumors, not only of endocrine origin expressed SST receptors, which even widened the list of neoplasms whose growth could be influenced via native hormone or its analogs. Besides pituitary adenoma and other neuroendocrine tumors, various brain tumors such as astrocytoma, medulloblastoma, oligodendroglioma, ganglioneuroblastoma, meningioma; and other cancers: renal cell, prostatic, breast, ovarian, colon, pancreatic and small cell lung cancer possess SST receptors.[15,16] Although several discrepancies concerning SST receptor expression exist in the literature, we should emphasize that even though the parenchymal cells of tumors did not express SST receptors as in the case of nonsmall cell lung cancer, they were often present in peritumoral veins, particularly in the smooth muscle cell layer, which seemed to play the role in tumor-host interaction.[17] Moreover, there were also data showing an antineoplastic effect of SST analogs on the tumors lacking SST receptors. In these cases SST and its analogs seem to act through indirect mechanisms such as the inhibition of the secretion and action of hormones and growth factors (e.g., IGF-I, EGF) involved in the growth processes of these tumors.

The cloning and functional characteristics of all 5 SST receptor subtypes (sst1-5) in 1990s led to the reevaluation of sst receptor subtypes presence in the tumoral tissues. Once again this analysis revealed that any tumor (endocrine or nonendocrine, primary or its metastasis) express one or more of sst receptor subtypes. Moreover, the density of sst receptors can differ significantly among various nonendocrine cancer cell types. Various tumors including gliomas, meningiomas, prostate, lung, pancreas and breast tumors express multiple sst receptor subtypes with sst2 being the most frequently expressed.[17] The majority of these tumors expressed also sst5 and, to a lesser extent, sst1,3 and sst4.[18] Attempt has been made to establish which subtypes of sst are involved in antineoplastic effect of SST analogs. Several studies indicate that not only sst2 but also other members of sst receptor family mediate antiproliferative action of SST and lead to cytostasis (cell cycle arrest) by the activation of the phosphotyrosine phosphatases and abrogation of various mitogenic signals, by the modulation of mitogen-activated protein kinases and subsequent induction of cyclin dependent kinase inhibitor p21, by the down-regulation of the phosphorylation of retinoblastoma Rb tumor suppressor gene and other mechanisms.[19-21] Besides cytostasis, SST analogs can also induce apoptosis of cancer cells acting mainly via sst3 and, to a lesser extent, via sst2 through p53 dependent and independent pathways and bax activation.[19,21]

It is worth recalling that whereas apoptosis is triggered at low SST agonist concentration (>0.1 nM), cytostasis is induced at much higher concentration (>50nM). Moreover, it is worth to emphasize, that the best anti-tumor effect of SST analogs is achieved when: (1) the initial tumor volume is small; (2) the continuous administration of SST analogs using osmotic minipumps is used (in comparison to subcutaneous injection); (3) higher doses of these drugs are applied; and (4) these drugs are used in combination with some other agents such as chemotherapeutic agents (taxol, 5-fluorouracil, doxorubicin), inhibitors of HMG CoA reductase, macrolides, tamoxifen and others what often leads to synergistic effects.[14]

The interest in SST and its analogs as the angiogenesis regulating factors began in the early 1990s when Reubi et al,[16] showed in pioneering studies that peritumoral veins located in the immediate vicinity of neoplasms expressed a high density of sst2. The antiangigenic activity of SST analogs was first demonstrated in 1991 by Woltering et al,[22] for octreotide and RC-160 in the chicken chorioallantoic membrane (CAM) model. Subsequent studies revealed that octreotide and other SST analogs exhibited antiangiogenic properties via the inhibition of endothelial cell proliferation, their adhesion, migration and invasion; via direct actions and also via indirect mechanisms such as inhibition of pro-angiogenic growth factors (VEGF, FGF, PDGF, IGF-I) secretion and action in several in vitro and in vivo model systems (CAM, Matrigel model, human umbilical chord vein endothelial cell (HUVEC) and others).[23] Moreover, in the recent years Watson et al,[24] have shown that sst2 gene is overexpressed only in proliferating angiogenic sprouts of human endothelial cells (angiogenic switch) but not in the quiescent endothelium. Such evidence suggests that sst2 may be a novel target for vascular delivery of anti-angiogenic SST agonists conjugated to cytotoxic agents or radioisotopes.

Other mechanisms through which SST analogs could evoke the anticancer effects is the regulation of the immune system. Several human lymphoid organs and cells contain SST and express various subtypes of sst receptors. The expression of sst2 is up-regulated following lymphocyte activation. SST analogs regulate several immune functions including lymphocyte proliferation, immunoglobulin synthesis and cytokine production.[25] However, most of these data concern mainly the in vitro effect of SST agonists.

Despite the large number of biological rationales concerning the control of cancer cell growth via SST analogs and the large number of preclinical studies investigating the anticancer effects of SST agonists, the use of currently available SST analogs in the common nonendocrine cancers (breast, colon, lung) generates much controversy. In a few clinical studies SST analogs were applied in the treatment of nonendocrine tumors: alone, as monotherapy in the patients with hepatocellular cancer;[26] or in combination with other treatment modalities, such as hormonal therapy for prostate and breast cancer,[27,28] and with chemotherapy for melanoma.[29] But the efficacy of SST analogs in patients with common nonendocrine cancers is rather disappointing. In some reports these drugs evoked higher response rate, longer symptom-free and survival time for the patients with cancers in comparison with conventional therapy. However, these encouraging results were not confirmed by other studies.

The discrepancies between disappointing clinical data and encouraging preclinical studies may depend on several factors such as the lack of sst2 receptors (the main target for currently available SST analogs) on human cancers (e.g., human pancreatic cancer cell lines); the predominance of other sst subtypes with low affinity to octreotide (e.g., sst1 in prostatic and ovarian cancers, and sst5 in colon cancer); too low doses of SST analogs applied in the patients (higher efficacy of OncoLAR than octreotide LAR); the desensitization of anticancer effect of SST analogs and others.

The development of new analogs both peptide and nonpeptide; subtype receptor specific, bispecific, universal (SOM 230) and chimeric compounds brings a new hope in the field of SST analogs therapy in patients with nonendocrine cancers. Some data coming from the studies concerning endocrine tumors such as no desensitization observed during SOM 230 administration in animal tumors encourage the investigators to examine the effects of these new compounds in the treatment of nonendocrine cancers.[30] Reubi et al,[31] have also described a peptide agonist KE108, which possesses extremely high affinity for all five sst receptor subtypes. Both these universal analogs may be a major advantage in the therapy of human tumors, which can express several sst subtypes concomitantly. Further progress in the treatment of cancers with SST analogs might to be linked with a new heptapeptide analog TT-232, currently undergoing the preclinical studies. TT-232, a heptapeptide with primary structure D-Phe-Cys-Tyr-D-Trp-Lys-Cys-Thr-NH2 has little effect on GH secretion but exerts a potent anti-tumor effect either in vitro on numerous cancer cell lines or on cancers grafted into immunosuppressed mice. This molecule binds to sst1 and sst5 receptors but it also is translocated to the cell nucleus and possibly acts via yet unknown intranuclear site.[32]

Although, the efficacy of currently available, "cold" SST analogs in the treatment of this kind of patients is scarce, their usefulness in the targeted radio- and chemotherapy of sst-positive nonendocrine tumors seems to be promising. Current knowledge and experiences in the field of sst-targeted radiotherapy are much more advanced than in the field of chemotherapy, with several preclinical and clinical trials concerning the efficacy of the former. The role of sst-targeted radiotherapy will be discussed in details in another chapter of this book. The progress in sst-targeted chemotherapy is slower and begins later on. However, the basis for the two modes of antitumor treatment is very similar. The wide spectrum of adverse reactions in patients with advanced cancers during conventional chemotherapy are caused by the sever toxicity of these agents to normal cells. Like sst-targeted radiotherapy, targeted chemotherapy is aimed at delivering the cytotoxic agents selectively and almost exclusively to tumor cells avoiding the exposure of several normal tissues. Schally and Nagy,[33] pioneered this concept with the development of the cytotoxic analogs of various hormones such as LHRH, bombesin, and somatostatin to treat LHRH receptor-, bombesin receptor- and somatostatin receptor-positive tumors, respectively. Schally and coworkers,[34,35] synthesized two different cytotoxic SST analogs, code-named AN-51, consisting of methotrexate linked to the SST analog RC-121 and AN-238, which is the RC-121 linked to a doxorubicin derivative. Both these compounds have intermediate binding affinities to sst-positive tissues in vitro suggesting that binding of cytotoxic agents changed the binding properties of SST analog.

In preclinical studies it was demonstrated that both of these compounds strongly inhibited tumor growth in many experimental mouse and rat models of human nonendocrine cancers such as breast, prostate, ovarian, colon, pancreatic, renal, endometrial and small and nonsmall cell lung cancer, glioblastoma (xenograft in animal model).[33-39] A much higher toxicity and lower or absent effectiveness on tumor growth was observed in animals treated with cytotoxic radical alone. In these studies the major side effect of sst-targeted chemotherapy was a transient fall in white blood cell count. In conclusion, sst-targeted chemotherapy is highly effective in preclinical tumor models and seems to be a promising approach to treat patients with sst-positive nonendocrine tumors.

SST Analogs in the Treatment of the Autoimmune/Chronic Inflammatory Disorders

Endogenous peptides SST and cortistatin, acting via sst receptors, modulate the immune system (see also Chapter 1). This finding constitutes a rationale for the attempts of therapeutic applications of SST analogs in the treatment of autoimmune and/or inflammatory disorders. The oldest indication of SST analogs for an autoimmune and inflammatory disorder is Graves' ophtalmopathy. The early studies with the application of 100 µg of octreotide three times daily during 3 months, resulted in approximatively 50% of improvements.[40,41] The more recent paper, concerning the administration of the modern form of lanreotide injected every two weeks, reports even better results, probably because of the better selection of patients.[42] In general, the effects are comparable with those obtainable by glucocorticoids. The therapeutic effect can be predicted by the positive scans showing the high expression of sst receptors within the orbital tissues. The treatment, in turn, resulted in a decrease of sst receptor scan positivity.[42] The mechanism of therapeutic effect of ocreotide and lanreotide in Graves' ophtalmopathy is complex. The inhibition of IGF-I activity, the suppression of cytokine production by the lymphoid cells and the direct effect on target cells within the orbit were listed as the possible explanations.[42] Summarizing, the treatment with octeotide and lanreotide should be recommended to the patients suffering from Graves' ophtalmopathy in the active phase of the disease and with positive sst receptor scans, who do not tolerate the glucocorticoids well. The high costs still constitute the main barrier of the therapy. Another condition which is considered as a target of SST analogs therapy is the rheumatoid arthritis.

The immunohistochemical studies on synovial biopsies from rheumatoid arthritis suffering patients have revealed the expression of sst2 in vascular endothelial cells of the synovial venules and on a subset of macrophages and fibroblast-like cells.[43,44] The expression of sst receptors was also demonstrated in affected joints by means of receptor scintigraphy.[45] It was shown that prolonged intra-atricular administration of SST-14 caused beneficial effects in patients suffering from rheumatoid arthritis.[46] The clinical trials with octreotide in the patients with rheumatoid arthritis refractory to usual anti-rheumatic drugs demonstrated the significant improvement.[47,48] Summing up, the long acting SST analogs of octreotide type seem to be useful in the treatment of rheumatic arthritis, although further clinical studies based on larger number of patients are needed.

SST Analogs in Diabetic Retinopathy and Senile Macular Degeneration

Proliferative retinopathies are the leading cause of blindness in the world and include diabetic retinopathy and age-related macular degeneration. In both subtypes of proliferative retinopathy the neovascularization is the main mechanisms involved in the retinal destruction, however, the pathogenesis of both forms is not fully understood. The sequence of events seems to be better known in diabetic retinopathy. Ischemic insult is thought to be the initiating factor causing changes in the existing microvasculature such as pericyte death and subsequent endothelial cell proliferation leading to the compensatory retinal neovascularization, which can adversely affect retinal function. Several experimental and clinical studies indicate that IGF-I and VEGF are two major players in the pathogenesis of these disorders. The rise of serum and vitreous IGF-I levels and vitreous VEGF level has been observed in patients with proliferative diabetic retinopathy.[49,50] Both factors have been shown to be the potent stimuli of ocular (retinal, iris, corneal and chorioidal) neoangiogenesis in several animal models in vivo and in vitro. Therefore, SST analogs which are known as the inhibitors of the secretion and action of IGF-I and VEGF could provide novel therapeutic opportunities to attenuate the improper angiogenic growth in the ocular tissues. Moreover, almost all sst receptor subtypes are expressed in the human eye with sst1 and sst2 being the most widely expressed in human retina.[51,52]

The first clinical trial assessing the therapeutic potential of SST analogs in the treatment of diabetic retinopathy were performed in the late 1980s. Octreotide is the most tested drug and was administered as the repeated subcutaneous injections or continuous subcutaneous infusions of octreotide, and as intramuscular injections of octreotide-LAR, hardly ever has another SST analog (BIM23014) been tested. In one of the first studies by Kirkegaard et al,[53] there was no observable beneficial effects of octreotide therapy in patients with diabetic retinopathy with exception of decreasing of GH/IGF-I levels and transient improvement of overall visual acuity of patients (only until cessation of therapy). In other studies various evidence of clinical improvement in the course of proliferative and nonproliferative diabetic retinopathy has been observed. The treatment with SST analogs resulted in the improvement in visual acuity, reduction of vascular leakage and the number, frequency and intensity of vitreous hemorrhage, slowing down the progression of retinopathy in the early stages of the disease and in the patients with advanced proliferative retinopathy.[54] After 3 years of octreotide therapy, in one study, the number of necessary vitreoretinal surgery was significantly reduced together with improvement in visual acuity and reduction of vitreous hemorrhages.[55] However, it should be underlined that each of these clinical trials was performed on a very small group of patients (1 to 23 patients per study), often without a control group, often with patients in various stages of the disease (early and advanced, proliferative and nonproliferative), and with a different period of time from another treatment regimen, often the duration of SST analogs therapy was too short as well as the follow-up period (from several days to several months, rarely 2-3 years). All these factors did not allow us to compare the obtained results and to draw the right conclusions about the effectiveness of SST analogs (both as prevention and therapy) in diabetic retinopathy. Even with these unreliable and mixed results, SST analogs remain the only therapeutic alternative to patients with proliferative retinopathy who have failed to respond to panretinal photocoagulation or cryotherapy, which both are invasive and painful procedures and sometimes are associated with some severe complications, such as reduced night and peripheral vision in patients.

Our experience in the treatment of age-related macular degeneration with SST analogs is even less examined. The pathogenesis of this disease which is the leading cause for blindness in elderly people, involved choroidal neovascularization, leakage of blood and serum and detachment of retinal pigment epithelium with scaring of the macula. Van Hagen et al,[51] reported that after 6 months of octreotide LAR treatment (20 mg monthly) 10 out of 15 eyes in 13 patients maintained their vision or even improved in visual acuity. A 2-year follow-up of these patients (all but one were treated for more than 2 years) showed that visual acuity remained stable. Similarly, in a pilot clinical study with lanreotide the stabilization of the disease progression was reported.

In conclusion, currently available data show that SST analogs therapy may retard progression of early and advanced diabetic retinopathy and delay the time required for laser surgery and in the case of age-related macular degeneration SST analogs may contribute to the recovery of function of retinal epithelium. However, a prospective randomized double blind placebo control trial is needed to assess reliably the effectiveness of these drugs in these diseases.

Conclusions

The growing knowledge on the incidence of somatostatin receptors and on their functions as well as the development of new molecules created the new possibilities of clinical application of SST analogs. The further progress could be expected in diagnosis and treatment of endocrine disorders, cancers and the autoimmune/inflammatory diseases with "classical" and/or novel SST analogs.

Acknowledgements

This review was supported by a grant no. 502-11-295 from Medical University of Lodz.

References

1. Hofland LJ, van der Hoek J, van Koetsveld PM et al. The novel somatostatin analog SOM230 is a potent inhibitor of growth hormone release by growth hormone- and prolactin secreting pituitary adenomas in vitro. J Clin Endocrinol Metab 2004; 89:1577-1585.
2. Rocheville M, Lange DC, Kumar U et al. Receptors for dopamine and somatostatin: Formation of hetro-oligomers with enhanced functional activity. Science 2000; 288:154-157.
3. Saveanu A, Lavaque E, Gunz C et al. Demonstration of enhanced potency of a chimeric somatostatin/dopamine molecule, BIM-23A387, in suppressing growth hormone and prolactin secretion from human pituitary somatotroph adenoma cells. J Clin Endocrinol Metab 2002; 87:5545-5552.
4. Warnet A, Harris AG, Renard E et al. Prospective multicenter study trial of octreotide in 24 patients with visual defects caused by nonfunctioning and gonadotropin-secreting pituitary adenomas. Neurosurgery 1997; 41:786-795.
5. Colao A, Filippella M, Di Somma C et al. Somatostatin analogs in treatment of nongrowth hormone secreting pituitary adenomas. Endocrine 2003; 20:279-283.
6. Hofland LJ, Lamberts SW. Somatostatin receptors in pituitary function, diagnosis and therapy. Front Horm Res 2004; 32:235-252.
7. van der Hoek J, Waaijers M, van Koetsveld PM et al. Distinct functional properties of native somatostatin receptor subtype 5 compared with subtype 2 in the regulation of ACTH release by corticotroph tumor cells. Am J Physiol Endocrinol Metab 2005; 289:E278-E287.
8. Pawlikowski M, Pisarek H, Kunert-Radek J et al. Immunohistochemical detection of somatostatin receptor subtypes in "clinically nonfunctioning" pituitary adenomas. Endocrine Pathol 2003; 14:231-238.
9. Zatelli MC, Piccin D, Bottoni A et al. Evidence for differential effects of selective somatostatin receptor subtype agonists on alpha-subunit and chromogranin A secretion and on cell viability in human nonfunctioning pituitary adenomas in vitro. J Clin Endocrinol Metab 2004; 89:5180-5188.
10. Gruszka A, Kunert-Radek J, Radek A et al. The effect of selective sst1, sst2, sst5 somatostatin receptors agonists, a somatostatin/dopamine (SST/DA) chimera and bromocriptine on the "clinically nonfunctioning" pituitary adenomas in vitro. Life Sci 2006; 78:689-693.
11. Pawlikowski M, Kunert-Radek J, Stepien H. Somatostatin inhibits the mitogenic effect of thyroliberin. Experientia 1978; 34:271-271.
12. Mascardo RN, Sherline P. Somatostatin inhibits rapid centrosomal separation and cell proliferation induced by epidermal growth factor. Endocrinology 1982; 111(4):1394-6.
13. Reubi JC. Octreotide and nonendocrine tumors: Basic knowledge and therapeutic potential. In: Scarpignato C, ed. Octreotide: From Basic Science to Clinical Medicine, Vol. 10. Basel: Karger, 1996:246-269.
14. Weckbecker G, Stolz B, Susini Ch et al. Antiproliferative somatostatin analogues with potential in oncology. In: Lamberts SWJ, ed. Octreotide: The Next Decade. Bristol: BiosSientifica Ltd., 1999:339-352.
15. Reubi JC, Laissue J, Krenning E et al. Somatostatin receptors in human cancer: Incidence, characteristics, functional correlates and clinical implications. J Steroid Biochem Molec Biol 1992; 43(1-3):27-35.
16. Reubi JC, Horisberger U, Laissue J. High density of somatostatin receptors in veins surrounding human cancer tissue: Role in tumor-host interaction? Int J Cancer 1994; 56(5):681-8.
17. Reubi JC, Schaer JC, Waser B et al. Expression and localization of somatostatin receptor SSTR1, SSTR2, and SSTR3 messenger RNAs in primary human tumors using in situ hybridization. Cancer Res 1994; 54(13):3455-9.
18. Patel YC. Somatostatin and its receptor family. Front Neuroendocrinol 1999; 20(3):157-98.
19. Ferjoux G, Bousquet C, Cordelier P et al. Signal transduction of somatostatin receptors negatively controlling cell proliferation. J Physiol Paris 2000; 94(3-4):205-10.
20. Sharma K, Patel YC, Srikant CB. C-terminal region of human somatostatin receptor 5 is required for induction of Rb and G1 cell cycle arrest. Mol Endocrinol 1999; 13(1):82-90.
21. Sharma K, Patel YC, Srikant CB. Subtype-selective induction of wild-type p53 and apoptosis, but not cell cycle arrest, by human somatostatin receptor 3. Mol Endocrinol 1996; 10(12):1688-96.
22. Woltering EA, Barrie R, O'Dorisio TM et al. Somatostatin analogues inhibit angiogenesis in the chick chorioallantoic membrane. J Surg Res 1991; 50(3):245-51.
23. Woltering EA. Development of targeted somatostatin-based antiangiogenic therapy: A review and future perspectives. Cancer Biother Radiopharm 2003; 18:601-609.

24. Watson JC, Balster DA, Gebhardt BM et al. Growing vascular endothelial cells express somatostatin subtype 2 receptors. Br J Cancer 2001; 85(2):266-72.
25. Ten Bokum AMC, Hofland LJ, van Hagen PM. Somatostatin and somatostatin receptors in the immune system: A review. Eur Cytokine Net 2000; 11:161-167.
26. Kouroumalis E, Skordilis P, Thermos K et al. Treatment of hepatocellular carcinoma with octreotide: A randomised controlled study. Gut 1998; 42:442-447.
27. Vainas G, Pasaitou V, Galaktidou G et al. The role of somatostatin analogues in complete antiandrogen treatment in patients with prostatic carcinoma. J Exp Clin Cancer Res 1997; 16:119-126.
28. Bontenbal M, Foekens JA, Lamberts SWJ et al. Feasibility, endocrine and anti-tumour effects of a triple endocrine therapy with tamoxifen, a somatostatin analogue and an antiprolactin in post menopausal breast cancer: Randomized study with long-term follow-up. Br J Cancer 1998; 77:115-122.
29. Medenica R, Janssens J, Caglayan et al. Successful treatment of metastatic melanoma with pulse therapy schedule of chemotherapy, interferon alpha, interferon gamma, interleukin-2 and octreotide. J Interferon Cytokine Res 1997; 17(suppl 20):S 109.
30. Weckbecker G, Briner U, Lewis I et al. SOM230: A new somatostatin peptidomimetic with potent inhibitory effects on the growth hormone/insulin-like growth factor-I axis in rats, primates, and dogs. Endocrinology 2002; 143(10):4123-30.
31. Reubi JC, Eisenwiener KP, Rink H et al. A new peptidic somatostatin agonist with high affinity to all five somatostatin receptors. Eur J Pharmacol 2002; 456(1-3):45-9.
32. Szende B, Keri GY. TT232: A somatostatin structural derivative as a potent antitumor drug candidate. Anticancer Drug 2003; 14:586-588.
33. Schally AV, Nagy A. Cancer chemotherapy based on targeting of cytotoxic peptide conjugates to their receptors on tumors. Eur J Endocrinol 1999; 141(1):1-14.
34. Nagy A, Schally AV, Halmos G et al. Synthesis and biological evaluation of cytotoxic analogs of somatostatin containing doxorubicin or its intensely potent derivative, 2-pyrrolinodoxorubicin. Proc Natl Acad Sci USA 1998; 95(4):1794-9.
35. Radulovic S, Nagy A, Szoke B et al. Cytotoxic analog of somatostatin containing methotrexate inhibits growth of MIA PaCa-2 human pancreatic cancer xenografts in nude mice. Cancer Lett 1992; 62(3):263-71.
36. Plonowski A, Schally AV, Nagy A et al. Inhibition of metastatic renal cell carcinomas expressing somatostatin receptors by a targeted cytotoxic analogue of somatostatin AN-238. Cancer Res 2000; 60(11):2996-3001.
37. Plonowski A, Schally AV, Koppan M et al. Inhibition of the UCI-107 human ovarian carcinoma cell line by a targeted cytotoxic analog of somatostatin, AN-238. Cancer 2001; 92(5):1168-76.
38. Kiaris H, Schally AV, Nagy A et al. Regression of U-87 MG human glioblastomas in nude mice after treatment with a cytotoxic somatostatin analog AN-238. Clin Cancer Res 2000; 6(2):709-17.
39. Kiaris H, Schally AV, Nagy A et al. A targeted cytotoxic somatostatin (SST) analogue, AN-238, inhibits the growth of H-69 small-cell lung carcinoma (SCLC) and H-157 nonSCLC in nude mice. Eur J Cancer 2001; 37(5):620-8.
40. Chang TC, Kao SC, Huang KM. Octreotide and Graves' ophtalmopathy and pretibial myxoedema. Brit Med Journal 1992; 304:158.
41. Krassas GE. Somatostatin anlogues in the treatment of thyroid eye disease. Thyroid 1998; 8:443-445.
42. Krassas GE. Ophthalmic Graves' disease: Clinical response to octreotide. In: Lamberts SWJ, Ghigo E, eds. The Expanding Role of Octreotide II: Advances in Endocrinology and Eye Diseases. Bristol: BioScientifica Ltd., 2002:115-128.
43. Ten Bokum AM, Melief MJ, Schonbrunn A et al. Immunohistochemical localization of somatostatin receptor sst2A in human rheumatoid synovium. J Rheumatol 1999; 26:532-535.
44. Takeba Y, Suzuki N, Takeno M et al. Modulation of synovial cell function by somatostatin in patients with rheumatoid arthritis. Arthritis Rheum 1997; 40:2128-2138.
45. van Hagen PM, Markusse HM, Lamberts SWJ et al. Somatostatin receptor imaging. The presence of somatostatin receptors in rheumatoid arthritis. Arthritis Rheum 1994; 37:1521-1527.
46. Coari G, Di-Franco M, Iagnocco A et al. Intra-articular somatostatin-14 reduced synovial thickness in rheumatoid arthritis: An ultrasonographic study. Int J Clin Pharmacol Res 1995; 15:27-32.
47. Paran D, Elkayam O, Mayo A et al. A pilot study of a long acting somatostatin analogue for the treatment of refractory rheumatoid arthritis. Ann Rheum Dis 2001; 60:888-891.
48. Koseoglu F, Koseoglu T. Long acting somatostatin analogue for the treatment of refractory RA. Ann Rheum Dis 2002; 61:573-574.

49. Grant MB, Caballero S, Millard WJ. Inhibition of IGF-I and b-FGF stimulated growth of human retinal endothelial cells by the somatostatin analogue, octreotide: A potential treatment for ocular neovascularization. Regul Pept 1993; 48(1-2):267-78.
50. Burgos R, Simo R, Audi L et al. Vitreous levels of vascular endothelial growth factor are not influenced by its serum concentrations in diabetic retinopathy. Diabetologia 1997; 40(9):1107-9.
51. van Hagen PM, Baarsma GS, Mooy CM et al. Somatostatin and somatostatin receptors in retinal diseases. Eur J Endocrinol 2000; 143(Suppl 1):S43-51.
52. Thermos K. Functional mapping of somatostatin receptors in the retina: A review. Vision Res 2003; 43(17):1805-15.
53. Kirkegaard C, Norgaard K, Snorgaard O et al. Effect of one year continuous subcutaneous infusion of a somatostatin analogue, octreotide, on early retinopathy, metabolic control and thyroid function in Type I (insulin-dependent) diabetes mellitus. Acta Endocrinol (Copenh) 1990; 122(6):766-72.
54. Grant MB, Mames RN, Fitzgerald C et al. The efficacy of octreotide in the therapy of severe nonproliferative and early proliferative diabetic retinopathy: A randomized controlled study. Diabetes Care 2000; 23(4):504-9.
55. Boehm BO, Lang GK, Jehle PM et al. Octreotide reduces vitreous hemorrhage and loss of visual acuity risk in patients with high-risk proliferative diabetic retinopathy. Horm Metab Res 2001; 33(5):300-6.

Index

T

V

Z